ANIMAL PRODUCTION AND HEALTH

ASSESSING QUALITY AND SAFETY OF ANIMAL FEEDS

Published by arrangement with the
Food and Agriculture Organization of
the United Nations
by
Daya Publishing House®
A Division of
Astral International Pvt. Ltd.
New Delhi - 110 002

"This book was originally published by the Food and Agriculture Organization of the United Nations (FAO) as FAO Animal Production and Health No. 160".

"The designations employed and the presentation of material in this publication do not imply the expression of any opinion, whatsoever on the part of the Food and Agriculture Organization of the United Nations concerning the legal status of any country, territory, city or area or of its authorities, or concerning the delimitations of its frontiers or boundaries."

'The designations 'developed' and 'developing' economies are intended for statistical convenience and do not necessarily express a judgment about the stage reached by a particular country, territory or area in the development process.'

First Indian Edition, 2007
Reprinted, 2016
ISBN: 978-81-7035-511-3

Published by : **Daya Publishing House®**
A Division of
Astral International Pvt. Ltd.
– ISO 9001:2008 Certified Company –
4760-61/23, Ansari Road, Darya Ganj
New Delhi-110 002
Ph. 011-43549197, 23278134
E-mail: info@astralint.com
Website: www.astralint.com

Printed at : **Salasar Imaging System,**
Delhi -110 035

CONTENTS

FOREWORD

Feeds and forages are variable in composition. Feed analysis provides information for farmers to optimize nutrient utilization in animal feeds; for feed compounders to prepare feed mixtures suitable for different animal production systems; for researchers to relate animal performance to feed characteristics; and for plant breeders to optimize the nutritive value of new varieties. Also of concern are the undesirable contaminants of animal feeds which may have a direct bearing on the safety of foods of animal origin. This book brings together six reviews on these subjects from the FAO Electronic Journal AGRIPPA in printed form.

The keynote article by Irene Mueller-Harvey describes current procedures for feed analysis and procedures to improve standards. She describes how to achieve quality control, quality assurance, laboratory accreditation and proficiency testing. Standard and widely accepted methods are described together with recent developments in feed analysis. Topics covered include: sample preparation, analysis of major components (dry matter, ash and minerals, crude protein, fat, fibres and starch) and of secondary plant products (tannins, mycotoxins and other contaminants). Developments in the analysis of whole samples by near infrared reflectance spectroscopy are mentioned and the potential of this technique to by-pass traditional feed analysis by directly predicting animal responses.

Feedstuffs vary because of genetic differences and as a consequence of feed processing. The paper by Gizzi and Givens considers the importance of these factors for the compound feed manufacturer, the farmer and the policy maker. Data variability also results from differences in the methodologies used to obtain the information. Chemical analysis procedures and animal study protocols may vary according to the laboratory or institute involved. Understanding the variation in chemical and nutritional characteristics of feedstuffs is vital to the effective use of feed information in livestock production.

The paper by Harinder Makkar describes the potential of the *in vitro* gas production method for evaluating nutritional quality of feed resources for ruminants. This technique enables selection of a feed or feed constituent for high efficiency of microbial protein synthesis in the rumen along with high dry matter digestibility, and provides a basis for development of feeding strategies to maximize efficiency. In addition, this technique provides an experimental tool to study the effects of various natural and synthetic compounds and their adverse or beneficial effects on rumen fermentation.

Felix D'Mello covers the microbiology of animal feeds, including forages, cereal grains, oilseed by-products and compound feeds. He notes the beneficial effects of lactic acid bacteria in the fermentation of forages during the process of ensilage. Lactic acid bacteria and yeast cultures have also been attributed with beneficial properties as feed probiotics for reducing scouring and increasing growth performance in farm animals. Animal feeds may become contaminated with harmful bacteria such as *Salmonella, Listeria* and *E. coli*. Cereal grains and oilseed by-products are regularly contaminated with fungi occurring as plant pathogens or developing during storage. Major adverse effects arise in farm animals due to the production of mycotoxins by certain species and strains of these fungi. He discusses potential methods for reducing the prevalence of deleterious fungi and regulations to control these feed contaminants, particularly mycotoxins.

In a second article, Dr D'Mello reviews the range of contaminants and toxins arising from anthropogenic and natural sources. He considers the distribution of heavy metals, radionuclides, mycotoxins, plant toxins, antibiotics and microbial pathogens in cereals, complete feeds and forages, together with the impact on farm livestock productivity and on the safety of resulting products. He considers methods of avoiding contamination and the regional significance of controls and legislation.

The final paper by Peter Hughes and John Heritage explores the developing controversy surrounding the use of antibiotics as growth promoters for food animals. These drugs are used at low doses in animal feeds and are considered to improve the quality of the product, with a lower percentage of fat and higher protein content in the meat. They may also help to control zoonotic pathogens such as *Salmonella, Campylobacter, Escherichia coli* and enterococci. Use of antibiotics, particularly at low doses, is associated with selection for resistance in pathogenic bacteria and it has been argued that the use of antibiotic growth-promoters may result in bacteria resistant to antibiotics used in clinical or veterinary practice, thus compromising the continued use of antimicrobial chemotherapy. The paper reviews the use of antibiotics as growth promoters and examines some of the alternative methods for achieving meat of high quality.

This publication intends to provide most recent information on the impact of animal feeds on food quality, food safety and the environment and to thus improve the basis for managing such risks which are increasingly at the centre of public and individual consumer attention.

Further articles will be published from time to time and can be read on-line at: http://www.fao.org/agrippa/ . Peer reviewed and edited documents are published in the AGRIPPA system and immediately available to readers. Subjects covered include livestock production, animal nutrition and feeding, and farming systems.

Samuel Jutzi
Director
FAO Animal Production and Health Division

LIST OF ABBREVIATIONS

ADF	acid detergent fibre
ATP	Adenosin-Three Phosphate
AAFCO	American Association of Feed Control Officials
AOAC	Association of Official Agricultural Chemists
BSE	bovine spongiform encephalopathy
CE	capillary electrophoresis
CF	crude fibre
CP	crude protein
CT	condensed tannins
CRM	certified reference materials
DAS	diacetoxyscirpenol
DDE	dichlorodiphenyldichloroethylene
DDT	dichlorodiphenyltrichloroethane
DM	dry matter
DOM	digestibility of organic matter
DON	deoxynivalenol
E-As	enzyme-linked antibodies
EE	ether extract
ELISA	enzyme linked immunosorbent assay
EU	European Union
FAPAS	Food Analysis Performance Assessment Scheme
FDA	Food and Drug Administration
GATT	General Agreement on Tariffs and Trade
GE	gross energy

GMO	genetically modified organisms
HPLC	high performance chromatography
HT	hydrolysable tannins
LAB	lactic acid bacteria
MALDI-TOF MS	matrix-assisted laser desorption/ionization time-of-flight mass spectrometry
ME	metabolizable energy
MIC	minimum inhibition concentration
mMBM	mammalian meat and bone meal
MRSA	methicillin-resistant *Staphylococcus aureus*
MS	mass spectrometry
NDF	neutral detergent fibre
NFE	nitrogen free extractive
NIR	near infrared reflectance spectroscopy
NIV	nivalenol
NOAH	National Office of Animal Health
OM	organic matter
OP	organophosphorus pesticides
OTA	Ochratoxin A
PCBs	polychlorinated biphenyls
PCR	Polymerase Chain Reaction
PDA	potato dextrose agar
PEG	polyethylene glycol
PTS	proficiency testing schemes
SAC	Scottish Agricultural College
SCFA	short cahin fatty acid
SMCO	S-methyl cysteine sulphoxide

VAM	Valid Analytical Measurement
VFA	volatile fatty acids
TLC	thin layer chromatography
TSE	transmissable spongiform encephalopathyes
vCDJ	variant Creutzfeldt-Jakob disease
VTEC	Vero-Toxigenic *Escherichia coli*
ZEN	zearalenone
WHO	World Health Organization

Modern techniques for feed analysis

I. Mueller-Harvey
Department of Agriculture, University of Reading
United Kingdom

This review describes accepted procedures for feed analysis with particular emphasis on rapid analyses. The analysis of individual chemical compounds tends to be easier to standardize than some less well-defined, but nutritionally important parameters, e.g. oil, fibre, starch. Inter-laboratory comparisons have demonstrated unacceptably large variation. A coordinated effort is required to reach consensus procedures amongst feed analysts using quality control measures. Unfortunately, there are few certified reference materials (CRMs) and proficiency schemes for animal feeds. In cases of complex mixtures of pro- or anti-nutrients, e.g. tannins, structure-function relationships have not yet been elucidated and, therefore, no particular assay can be recommended to plant breeders. Finally, the concept of 'traceability of analytical measurements' has been translated to 'traceability of feeds' in the form of a 'Feed Passport'. This will become increasingly important following the meat-and-bone meal issue in cattle feed and new soybean varieties that led to the bovine spongiform encephalopathy (BSE) crisis and the genetically modified organisms (GMOs) disputes.

1. INTRODUCTION

Most forage and by-product feeds are variable in composition. Only very few, such as dried whole whey, are uniform in composition and do not usually require analysis (Schingoethe, 1991). Feed analysis provides information for:

- farmers to optimize nutrient utilization in animal feeds;
- feed compounders to prepare feed mixtures suitable for different animal production systems;

- researchers to relate animal performance to feed characteristics; and
- plant breeders to optimize the nutritive value of new varieties (Madsen *et al.* 1997; Wrigley, 1999).

Many rapid analytical tests have been developed over the last few years and are of use when faced with a need for quick decisions or when confronted by large numbers of samples, e.g. at entry points of shipments, trading situations, on contamination sites or in plant breeding programmes.

This review will cover the main issues which prompted developments in quality assurance and control of analysis. Information will also be given on accreditation of laboratories, together with useful addresses. Subsequent sections will describe standard and widely accepted methods, highlight areas that require particular attention and refer to recent developments in feed analysis. Topics covered will include: sample preparation, analysis of major components (dry matter, ash and minerals, crude protein, fat, fibers and starch) and of secondary plant products (tannins, mycotoxins and other contaminants). Developments in the analysis of whole samples by near infrared reflectance spectroscopy will be mentioned and the potential of this technique to by-pass traditional feed analysis by directly predicting animal responses.

The reader may consult the following books for general information on feedstuff evaluation and diet formulation in relation to animal production: Wiseman and Cole (1990), Minson (1990), AFRC (1993), Fahey *et al.* (1994), Givens *et al.* (2000a). Detailed protocols for analytical methods can be found in MAFF (1986), Watson (1994), AOAC (1995) and on the web (Appendix 1).

This review will only briefly mention a few chromatographic and mass spectrometric techniques in relation to analysis of secondary products. It will not cover the analysis of whole samples by new *in vitro* digestibility techniques, pyrolysis-mass spectrometry, nuclear magnetic resonance or other spectroscopic techniques as these have been covered elsewhere (Mauricio *et al.*, 1999; Givens *et al.* 2000a).

2. QUALITY CONTROL, ASSURANCE, ACCREDITATION AND PROFICIENCY TESTING IN FEED ANALYSIS

2.1 Inter-laboratory variation

The need for agreement on methods to obtain comparable, useful data became obvious with the beginning of 'scientific agriculture' in the 19th century (Midkiff, 1984). Different methods led to widely varying results and to regulatory confusion. In 1884, the first meeting of the Association of Official Agricultural Chemists (AOAC) began to tackle fertilizer analysis and soon, 1886, included feedstuff testing. The AOAC president stated in 1896 'The matter of the analysis of foods and feedstuffs, as shown by the experience of the association, is one of the most difficult questions connected with the work of this organization'. The AOAC was renamed in 1965 to 'Association of Official Analytical Chemists' to take cognizance of the fact that the AOAC now had a wider scope (Midkiff, 1984).

A report in 1985 on the quality of data relating to Pb and Cd analysis in food laboratories concluded that results were inaccurate and the validity of the data was in doubt (Patey, 1996). Similar conclusions have been reached for a wide range of data relating to feed analysis. It would appear that feed analysis is somewhat lagging behind food analysis. Selected examples are given below, in order to illustrate the magnitude of the problem.

Bailey and Henderson (1990) concluded that there was an urgent need to improve oil and sugar determinations, since these methods had relatively poor precision amongst 15 feed laboratories. These data are commercially important as they are used when making labeling declarations and for energy content estimation. Lanari *et al.* (1991) commissioned a study of brewer's grain, dried beet pulp, lucerne and hay with 20 feed analysis laboratories. Unacceptably large coefficients of variation were found for mean oil content (determined as ether extract): 3.5 percent (cv = 17.8); lignin: 5.4 percent (cv = 27.3); neutral detergent fibre (NDF) 55.6 percent (cv = 7.4 percent). Coefficients of variation were 5 percent or less for dry matter (DM), crude protein (CP), crude fibre (CF), ash, acid detergent fibre (ADF), and gross energy (GE). Beever *et al.* (1996) submitted two contrasting maize silages to 10 commercial feed laboratories. They concluded that current feedstuff analysis provided unacceptable variation and required national standards, as coefficients of variation were 12.7 percent for CP and 16 percent for starch. In addition, Givens *et al.* (2000b) also list several more references reporting large inter-laboratory variation for several analyses relevant in feed evaluation: organic matter (OM) digestibility, *in situ* rumen degradation, gas production kinetics, *in vitro* digestibility, metabolizable energy (ME) and GE.

In vitro and *in vivo* digestibility measurements have been shown to suffer from similar inter-laboratory variation. Madsen and Hvelplund (1994) sent five feeds to 23 laboratories in 17 countries and found "differences ... in protein degradabilities between laboratories too large to be acceptable" when using nylon bags in the rumen of cows or sheep. The within laboratory variation, however, of protein degradation was acceptable. They concluded that sample preparation and processing and the bags themselves varied considerably between laboratories and made detailed recommendations for the nylon bag procedure. Authors also recommend that a 'standard feed is made available for all laboratories for routine checking of analytical procedure'. A small ring test of the Tilley-Terry method between 3 laboratories suggested that it was a robust method for OM digestibility of roughages (Madsen *et al.*, 1997). However, problems were encountered with concentrates and highlights that careful standardization is required. Not surprisingly, feed manufacturers have moved to an enzyme method and such results are expected to be more reproducible (Madsen *et al.*, 1997).

There are large differences between the different *in vitro* digestibility techniques (rumen fluid versus cellulase based techniques) which are used to predict *in vivo* digestibility (De Boever *et al.*, 1994; Aufrère and Michalet-Doreau, 1988). Clearly, such differences in methods need to be resolved.

These data collectively support the introductory statement that feed analysis can be difficult. Strict adherence to method details is important, especially if the method itself defines the component, e.g.

crude fat, NDF, lignin, as empirical fractions are chemically not well defined (Bailey and Henderson, 1990). Zeeman and Bonn (1995) have also suggested that there should be an international definition of starch. However, large variation may highlight problems in analytical methods and thus encourage closer investigation. Problems with too high NDF values, for example, are likely to be stem from incomplete solubilization of starch. Recently, Thiex *et al.* (1996) investigated the observed large variation in reported vitamin A results, which was noted by the American Association of Feed Control Officials (AAFCO) who operate a feed check sample programme. Several recommendations were made to reduce errors of Vitamin A in animal feed and pet food analysis.

2.2 The benefits of quality assurance programmes

A review of the literature for methods of feed grain analysis concluded that it was not possible to assess if variation in reported values was due to genotypic, environmental factors or inter-laboratory differences. Petterson *et al.* (1999), therefore, recommended the use of quality assurance schemes, inter-laboratory evaluation programmes and reference materials.

The Global Environmental Monitoring Scheme of the World Health Organization (WHO) tested the performance of European Union (EU) laboratories that contribute data on food contamination (Weigert *et al.*, 1997). This involved five proficiency tests, 136 laboratories in 21 countries using their own preferred methods for the analysis of trace metals (Pb, Cd, Hg in milk powder), pesticides (organochlorine, organophosphorus, pyrethroid in spinach powder), nitrate in spinach powder, aflatoxins in nut-based animal feeds. Only 60 percent reported accurate for trace metals, 41 percent for pesticides, 43 percent for nitrate, 88 percent for aflatoxins and 53 percent for patulin (average = 68 percent).

Key *et al.* (1997) summarized the results of the UK Food Analysis Performance Assessment Scheme (FAPAS) from 1990-1996. For pig feeds (moisture, ash, oil, protein, fibre, Cu), only 76 percent of laboratories achieved satisfactory results and for nutritional analysis

80 percent were satisfactory. The FAPAS study also reveals that some analyses are more difficult than others. 91 percent of aflatoxin data were satisfactory, 81 percent of veterinary drug residues, 86 percent of dichlorodiphenyldichloroethylene (DDE), but only 71 percent of dichlorodiphenyltrichloroethane (DDT) and 65 percent of Ca data. Surprisingly, Ca analysis has shown little improvement over the years. Inter-laboratory comparisons and proficiency testing can highlight inappropriate methods, e.g. analysis of veterinary drug residues using certain immunological tests. However, once laboratories were participating in proficiency tests on a regular basis (e.g. FAPAS) the average percentage for accurate results increased.

Horwitz (1993) observed "most experimentation dealing with analytical methodology in biological sciences has been conducted within a single laboratory. Method validation by other laboratories was considered not only unnecessary but also detrimental because, in the words of one commentator, 'the results are too variable'. Within the last two decades, however, it has become increasingly apparent that a collaborative inter-laboratory study is the only way to estimate the variability characteristics of methods...." and to meet the increasing demand by regulatory programmes for high quality data.

2.3 How to achieve valid data

The UK Department for Trade and Industry launched an initiative in 1994 on Valid Analytical Measurement (VAM) incorporating four main principles.

2.3.1 Principle 1: Measurements should be made using properly validated methods

Properly validated methods will provide information on the performance of an analytical technique, such as accuracy and precision, ruggedness, operating range, selectivity and limits of detection. It is essential when reporting a measured value to also give its uncertainty. Otherwise, it is not possible for users of the data to know what confidence to place in the data. It needs to be recognized that problems of communication can arise from the word 'uncertainty', as a layperson may misinterpret the statistical term 'error' into 'inaccurate data'! (Williams, 1996).

Guidelines are now available for the statistical evaluation of analytical tests and laboratory performance (Bailey and Henderson, 1990): the inter-laboratory precisions are generally a function of concentration. Horwitz' group found that the within-laboratory variation was approximately one-half to two-thirds of the between-laboratory variation and can be used as a 'bench mark for judging previously unevaluated methods' (Bailey and Henderson, 1990).

2.3.2 *Principle 2: Quality assurance protocols should incorporate certified reference materials (CRMs) to ensure traceability of measurements*

CRMs are used for:

- Calibration and verification of measurements under routine analysis conditions
- Internal quality control and quality assurance schemes
- Verification of the correct application of standardized methods
- Development and validation of new methods of measurement.

Unfortunately, only few CRMs are available for validation of analytical methods for feed analysis, especially for proximate analysis (Crosby, 1995). However, CRMs exist for minerals in two animal feeds: hay powder (Ca, K, Mg, P, S, Zn, I, N, Kjeldahl-N) and rye grass (As, B, Cd, Cu, Hg, Mn, Mo, Ni, Pb, Sb, Se, Zn) (Maier *et al.*, 1990; see Appendix 1 for CRM suppliers). Information on the in-house production of reference materials can be found in Walker and Brookman (1998). It is good practice to include external CRMs or in-house reference materials into all analytical procedures.

2.3.3 *Principle 3: Laboratories should seek an independent assessment of their performance for particular tests, preferably by participating in national and international proficiency testing schemes (PTS)*

PTS independently assess the performance of analytical laboratories. The true concentration of an analysis can be determined by addition of a known amount of analyses to a base material or better still through the use of a consensus value produced by a group of analysts. Guidelines exist for organizing a collaborative study to evaluate analytical methods (AOAC, 1988). Appendix 1 lists several proficiency testing schemes.

2.3.4 *Principle 4: Laboratories should seek independent approval of their quality assurance arrangements, preferably by accreditation or licensing to a recognized quality standard.*

It should be recognized that only certain accreditation schemes are appropriate for laboratories performing chemical analysis, e.g. NAMAS M10, ISO 90025, ISO/IEC 17025, EN 4501. General guidelines have been written on how to prepare for accreditation (EURACHEM/WELAC, 1998). Gangaiya and Morrison (1992) listed general problems of setting up quality assurance in some developing country laboratories. Appendix 1 lists several accreditation organizations.

3. SAMPLE PREPARATION

Good analytical data require that samples be representative of the whole and that their integrity has been ensured during transport to the laboratory and during their preparation (drying and grinding). Relevant guidelines for feeds can be found in Feeding Stuffs (1988), AOAC (1995), Crosby (1995) and Wrigley (1999). Drying may adversely affect the analysis of sugars, vitamins, certain trace elements (F, Se, B), and ammonia and volatile fatty acids (VFAs) in silages (MAFF, 1986). Caution needs to be taken when grinding lupin and chickpea seeds (Petterson *et al.*, 1999).

4. MOISTURE

Several methods exist to determine moisture content of feeds: oven drying at 105° C for 16h, 125° C for 4h or 135° C for 3h (AOAC, 1995). Molasses should be dried at 70° C because of high levels of volatile compounds (Petterson *et al.*, 1999). Oven-drying is problematic with silages and high fat feeds; vacuum-oven drying at 95-100° C (AOAC,

1995) or at <70° C, Karl Fischer or toluene distillation are alternative techniques (Crosby, 1995; Cherney, 2000). Baker *et al.* (1994) found significant variation in moisture contents between laboratories, which were tracked down to variable temperature gradients in ovens (Givens, pers. communication).

5. ASH

Crude ash is determined either by ashing at 600° C for 2 h or between 500-550° C for 12-16 h (Midkiff, 1984; Petterson *et al.*, 1999). For difficult samples, Crosby (1995) lists some special techniques. Flameless atomic absorption spectrometry on solid feed samples for trace levels of Cu represents an interesting development (Anzano *et al.*, 1994). However, sample size at present is restricted to 2-4 mg. This technique, therefore, requires further development for general applicability in feed analysis.

6. CRUDE PROTEIN

Crude protein data are standard for evaluating the protein value of forages (Cherney, 2000). The historic developments and critical points in Kjeldahl nitrogen measurements (choice of catalyst, temperature and digestion times) have been summarized by Lakin (1978) and Midkiff (1984). The Dumas technique represents an alternative method for total nitrogen. Total nitrogen is determined after combustion of the sample and several commercial instruments are now available for this. The measurements are rapid taking only 2-5 minutes per sample. The small sample size (20 to 500 mg) is the key problem with the Dumas method. Samples should be finely ground and preferably dry to avoid too frequent changes of costly reagents. Therefore, the best method for some materials, e.g. silages, is still the Kjeldahl procedure as drying would remove ammonia-N.

The need for corrosive and toxic reagents is the main disadvantages of the Kjeldahl method.

Total Dumas nitrogen values can be slightly higher than the Kjeldahl values as Dumas N also includes nitrate and organic compounds that are highly resistant to acid digestion (Lakin, 1978;

Petterson *et al.*, 1999). As a result, exceptional differences can occur with some biological matrices, e.g. fruits, vegetables and fish, with Kjeldahl-N:Dumas-N ratios as low as 0.15 (Simonne *et al.*, 1997).

A factor of 6.25 is used to convert total nitrogen in animal feeds into crude protein. However, the amino acid composition varies between foods and therefore different factors have been suggested to convert total N to crude protein (Petterson *et al.*, 1999). It can range from 5.14 for grains and oilseeds to 6.38 for dairy products. As there is some doubt about the universal validity of such factors, authors should report crude protein values together with the factors used in their calculation (Lakin, 1978).

It should be noted that none of the dye-binding methods were universally acceptable between laboratories for protein determinations (Petterson *et al.*, 1999). Recently, Strong and Duarte (1992) described a simple and rapid biuret method for protein determination in wheat, rice and soybeans. This has been applied to a wide range of different grains and only requires a blender, reagents, colorimeter at 550 nm and minimum operator training. The method requires initial calibration against protein values obtained by other techniques. This procedure may be of interest to grain and feed merchants as it can be completed in five minutes.

7. FIBRE ANALYSIS

It is important to recognize that all fibre determination employs 'empirical' methods, i.e. the method determines the final result and any deviation from the analytical protocol will produce a different result (Midkiff, 1984; Crosby, 1995). Midkiff (1984) summarized the history of crude fibre analysis. Cherney (2000) contrasted the Weende proximate analysis system, which originated in 19th century and has hardly changed since then, with the Van Soest system developed in the 1960s. He cautions against the use of CF, nitrogen free extractives (NFE), ether extract (EE) in feed evaluation systems, as they do not sufficiently separate digestible from non-digestible fractions. The Van Soest system is now widely used for forage evaluation as it provides useful measurements for nutritionally important parameters, such as structural carbohydrates (Goering and Van Soest, 1970; AOAC method 973.18).

Several groups have since modified the original manual method (Cherney *et al.*, 1989; Van Soest *et al.*, 1991) and Chai and Uden (1998) have described a simple oven-based procedure. The micro-NDF method by Pell and Schofield (1993) is another modification, which allows NDF to be determined on small sample sizes (10 to 50 mg). Moore and Hatfield (1994) comprehensively reviewed the composition and analytical methods for structural and non-structural carbohydrates citing research involving both ruminants and monogastric animals. Mertens (1997) concluded, "the only fibre method that can be used on all types of feeds is the method recommended by the National Forage Testing Association" (Undersander *et al.*, 1993, Hintz and Mertens, 1996; see also Appendix 1 for Web site).

Starch removal from NDF can be difficult and requires pre-treatment with α-amylase either overnight (McQueen and Nicholson, 1979) or by using heat-stable α-amylase (Sigma product A3306) during the last 30 minutes of NDF extraction and during filtration (Cherney *et al.*, 1989). Difficult samples may also require pretreatment with 8M urea (Van Soest *et al.*, 1991). Filtration problems may be due to incomplete starch removal. However, errors in fibre analysis may also stem from aged crucibles, as crucibles can be damaged at too high temperatures or by rapid rates of heating or cooling in muffle ovens (Crosby, 1995).

Fibretec (Tecator, Hoenganaes, Sweden) and the recent FibreAnalyzer (ANKOM Technology Co., Fairport, NY, USA) are instruments for NDF and ADF extractions. Studies indicated that ADF by Fibretec and FibreAnalyzer are comparable for most feeds. However, in our experience NDF in starch containing feeds is best analysed by the FibreAnalyzer as it removes starch more efficiently (unpublished data).

8. STARCH

Starch consists essentially of two components, amylose and amylopectin. Amylose polymers contain up to 2000 glucose units connected through linear 1-->4 linkages. Amylopectin is a highly branched polymer containing 2,000 to 220,000 glucose units with 1-->4 and 1-->6 linkages. Therefore proper extraction techniques are crucial for successful starch analysis (Petterson *et al.*, 1999).

Starch is determined by pre-treating with 80 percent ethanol in boiling water to remove low molecular weight sugars, followed by gelatinization and solubilization, before extracting and hydrolyzing the starch (Faichney and White, 1983; Åman and Graham, 1990; Hall, 1997). Incomplete dissolution and incomplete accessibility to enzymes tend to be the main problems in starch analysis. McCleary *et al.* (1994; 1997) reported that a ringtest involving 29 laboratories to evaluate the Megazyme enzyme kit for cereal products and some animal feeds produced good agreement for total starch content. This enzyme kit was accepted as an AOAC method (996.11) (Petterson *et al.*, 1999).

The proportion of amylose and amylopectin depends on the source of the starch. Starch digestibility is not necessarily related to total starch content as the amylose: amylopectin ratio and processing affect the extent of starch digestion (Allen *et al.*, 1997; Reynolds *et al.*, 1997). The Rapid Visco Analyser measures pasting properties of starch in grain and Wrigley (1999) speculated that this may be related to starch digestibility, but this awaits further research.

9. CRUDE FAT

Fat in plant-derived feeds consists mainly of mono-, di- and triacylglycerides, free fatty acids and phospholipids. Feeds also often contain fats from animal and other waste products. Depending on the rendering processes used, heating or storing can lead to unsaponifiable matter, oxidized and polymerized fatty acids which will contribute to crude fat values, but not be of nutritional value (Edmunds, 1990). Midkiff (1984) described the history of crude fat analysis and the sample types that caused problems.

Several official methods exist to determine crude fat in animal feeds (AOAC 945.16 and 920.38; MAFF 1986). Crude fat methods are empirical methods and procedural details must be closely adhered to. They are based on solvent extractions with or without hydrolysis. Crude fat is extracted by the EU procedure A with petroleum ether (40-60° C) and the dried residue weighed (Feeding Stuffs 1988). Acid hydrolysis is used as a pretreatment in EU procedure B. However, it is recommended that animal feeds are first extracted with petroleum ether (40-60° C) (Procedure A) and then subjected to acid hydrolysis, before re-extraction with petroleum ether (40-60° C) (Procedure B). This minimizes losses during filtration of the acid digest, which can occur if protected fats, i.e. Ca and Mg salts of fatty acids are present (Edmunds, 1990; Crosby, 1995).

Neither procedure A nor B will give satisfactory results if feeds contain milk products. The Rose-Gottlieb method is required for such feeds, as the alkaline pretreatment frees occluded lipids from protein capsules (ISO 1211:1984). Similarly, canned dog foods require extraction with a series of solvents after hydrolysis (Budde, 1952; Midkiff, 1984; AOAC 954.02).

Problems may arise if diethyl ether is used instead of petroleum ether as suggested in some methods. Slightly higher values may result if water has not been removed completely from either the sample or the solvent, as some compounds, such as urea and sugars, are slightly soluble in diethyl ether in the presence of small amounts of water (Midkiff, 1984; AOAC 920.39).

Great care is required when performing solvent extractions. Solvent are usually recycled between a lower electric heating source and an upper water-cooling system in a Soxhlet apparatus. Accidents are not uncommon with such a set-up. For this reason, a different type of extractor has been developed recently (Brown and Mueller-Harvey, 1999). The Soxflo instrument requires neither heat nor cooling water and is based on a dry-column procedure. The sample is packed into a column and the extracting solvent drips slowly through it. Crude fat values determined by the Soxflo and Soxhlet procedures were found to be in close agreement. If sample drying and packing procedures are evaluated against the Soxhlet procedure, good and reliable crude fat data can be obtained in 60 to 90 minutes.

10. SECONDARY PLANT PRODUCTS: TANNINS

Although a whole range of different plant secondary products exists, only tannins will be discussed here. The main reason for this is the persistent confusion that surrounds tannins in animal nutrition. Analytical techniques have often been misapplied and data misinterpreted. However, reference will be made to rapid analyses of mycotoxins and other secondary plant products which can contaminate or affected the nutritive value of animal feeds (see Section 11).

Tannins comprise a diverse group of phenolic compounds, varying in molecular size from 500 to possibly 28,000 Daltons (Jones *et al.*, 1976; Mueller-Harvey and McAllan, 1992; Schofield *et al.*, 2001). Some are easily extracted by aqueous solvents, others are not and can be measured as fiber or protein bound tannins (Jackson *et al.*, 1996; Reed, 1986). The naming and thereby classification of condensed (CT) or hydrolysable tannins (HT) is somewhat misleading as some of the 'condensed' tannins are relatively easily degraded oxidatively, whereas some of the 'hydrolysable' tannins will resist all attempts at hydrolysis (Mueller-Harvey, 1999; 2001).

Negative and positive animal responses have been attributed to tannins, i.e. ranging from animal death to increased growth rates (Butter *et al.*, 1999; Mueller-Harvey, 1999). Despite many animal studies involving tanniniferous feeds, little attempt has been made to elucidate the relationship between animal production and tannin structures. Progress is unlikely to be achieved by the continued use of colorimetric tannin assays alone (Petterson *et al.*, 1999). Lowry *et al.* (1996) put it succinctly: "the simplicity of absorbance measurements masks the problems of extracting meaningful data".

A general reagent produces varying colour yields from different phenolic compounds, which includes tannins (Folin and Ciocalteau, 1927). However, phenolics and tannins tend to occur in mixtures and quantitation is not possible unless isolated standards are used, which are representative of the exact composition of the material being examined. Procedures for colorimetric methods for CT and HT have been described by Graham (1992), Waterman and Mole (1994) and Hagerman (Web site).

Porter *et al.* (1986) recommended a modification of the butanol-HCl assay. This assay is excellent for detecting the presence of CT. However, for quantization purposes it should only used if isolated tannins are used as calibration standards as tannin structures have a marked effect on colour yield (Giner-Chavez *et al.*, 1997). Even within the same species, there can be sufficient structural variation to warrant isolated tannin standards from different accessions in quantitative work (Stewart *et al.*, 1999). More recently, a CT assay based on acidified 4-dimethylamino-cinnamaldehyde (DMACA-HCl) was described (Li *et al.*, 1996). This assay proved more sensitive than the vanillin-HCl method, which can suffer from interference by water (Terrill *et al.*, 1990). The DMACA assay was also used successfully on thin layer chromatography (TLC) plates and as a histochemical assay. This method has not yet been examined thoroughly against interferences from monomeric flavanols.

The analysis of HT has been reviewed by Mueller-Harvey (2001). Again, assays are best used to detect HT, as accurate quantitation would require isolated standards (Hagerman *et al.*, 1997). Furthermore, there are just a few known species that produce only gallotannins, from which gallic acid can be hydrolysed before detection (Willis and Allen, 1998). More usually, gallo- and ellagitannins occur together as mixtures. Ellagitannins are much more difficult to measure (Wilson and Hagerman, 1990) although the free ellagic acid can be measured by high performance chromatography (HPLC) (Mueller-Harvey *et al.*, 1987). However, as stated above, many other so-called HT does not release gallic or ellagic acid.

It is recommended that additional assays be used in tannin analysis to overcome some of the problems of the colorimetric tests, e.g. the Yb-precipitation method (Reed *et al.*, 1985; Giner-Chavez *et al.*, 1997; Krueger *et al.*, 2000a) and/or one of the protein or polymer binding assays (Makkar *et al.*, 1987; Dawra *et al.*, 1988; Makkar *et al.*, 1993) when screening novel fodder plants. If facilities are available to measure radioactivity, tannins can also be measured by binding to 125I labelled BSA (Hagerman and Butler, 1980) or 14C-polyethylene glycol (PEG) (Silanikove *et al.*, 1996). Tannins can be estimated directly in feed samples, without prior extraction in the latter method.

The Yb-method precipitates total phenolics (i.e. tannins, flavonoids and other phenolics) which correlate positively with butanol-HCl CT (Reed *et al.*, 2000). The advantage is that phenolics are determined by gravimetry without the need for standards. Interestingly, the amount of PEG bound per gram of sample correlated highly significantly with *in vitro* N digestibility and appeared to be a more meaningful measurement than the colorimetric asssays (Jones and Palmer, 2000). Correlations with CT measurements by vanillin-HCl were better than by butanol-HCl.

In view of the problems of colorimetric assay, it seems surprising that TLC has not been used more widely. TLC reveals the presence of CT vs HT, low vs high molecular weights of CT, the subclasses of CT or HT (procyanidins or prodelphinidins, gallo- or ellagitannins), plus semiquantitative information based on colour intensity (Mueller-Harvey *et al.*, 1987). Tannins from different species have also been compared by reverse phase HPLC (Mueller-Harvey *et al.*, 1987), size exclusion chromatography (Yanagida *et al.*, 1999; Hedqvist *et al.*, 2000) and normal phase chromatography (Tanaka *et al.*, 1984; Okuda *et al.*, 1989; Hagerman *et al.*, 1992). Okuda *et al.* (1989) reviewed the mass spectrometric analysis of tannins. More recently, ESI-MS (Guyot *et al.*, 1997; Marais *et al.*, 2000) was used to determine tannin molecular weights. New developments in MALDI-TOF mass spectrometry (Hedqvist *et al.*, 2000; Krueger *et al.*, 2000a and b) succeeded in determining the molecular weights and composition of complex tannin mixtures.

11. IMMUNOASSAY TECHNIQUES: FOR MYCOTOXIN AND OTHER FEED CONTAMINANTS

Mycotoxins pose severe hazards to humans and animals. Mycotoxins in feeds, along with many other organic compounds, can be analysed by either instrumental [HPCL, gas chromatography (GC), capillary electrophoresis (CE)] or biochemical (immunoassays) techniques.

Skerritt and Appels (1995) described the basic principles of enzyme-linked immunoasorbent assays (ELISAs) and illustrated the different forms of techniques (direct or indirect assays, sandwich or competition assays). For example, in the direct competitive ELISA format, enzyme-linked antibodies (E-As) coat the surface of a well plate or test tube, which are supplied in a test kit. The operator extracts the feed sample and places the extract onto the surface. Compounds of interest, i.e. antigens, bind to the E-As thereby releasing them from the surface. E-As are then washed away and an enzyme substrate is added to the remaining, bound E-As. This procedure results in an inverse relationship between antigen concentration and colour production: the higher the antigen concentration, the more E-As have been washed away and the smaller the colour yield of the reaction.

Several commercial kits are available for aflatoxins, zearalenone and other mycotoxins, alkaloids, glucosinolates, insecticides, herbicides, fungicides, various environmental pollutants, vitamins in foods and animal feeds. Morgan (1995) reviewed these ELISA techniques and listed suppliers' addresses. Schneider *et al.* (1991) developed an interesting dipstick technique for the simultaneous detection of several mycotoxins.

ELISAs can be used by regulatory authorities, quality control laboratories and in research laboratories. They require relatively little user training and can be used in small laboratories or under field conditions. Low cost and high speed make ELISAs ideal for on-site monitoring of stored grain and for assessing rapidly if the maximum residue levels of traded animal feeds have been exceeded. They can also be used for checking suspected spillages. High sample throughput facilitates elimination of large numbers of negative, uncontaminated samples. It is, however, recommended that samples giving positive results for a contaminant should be re-analysed by conventional instrumental techniques to ensure absence of matrix interference and accuracy of the data (Petterson *et al.*, 1999).

Some commercial test kits compare well with standard AOAC methods for aflatoxins in animal feeds (Trucksess *et al.*, 1989 and 1990; Cochrane, 1991) and can be used for quantitative or semiquantitative measurements in as little as 3 minutes. Several of these have now obtained AOAC approval (Trucksess *et al.*, 1989). ELISAs can be tailored to be selective for individual compounds or compound classes, parent compounds or metabolites and even for isomers. Highly sensitive ELISAs for M1 in milk samples have been reported (Kawamura *et al.*, 1994). Of interest in the current European BSE crisis is a report by von Holst *et al.* (2000), who evaluated the applicability of a commercial ELISA method to detect proper heat treatment of pork and beef meals.

A word of caution. ELISA procedures – like any analytical method – need to be carefully evaluated as some solvents used for analyte extraction may adversely affect antibody performance (Morgan, 1995). In an inter-laboratory study, ELISA tests gave good results for ZON, but accuracy with the DON kits was poor (Schuhmacher *et al.*, 1997). Matthews *et al.* (1996) investigated commercial kits for three organophosphorus pesticides (OP) for laboratory and field use. The authors found good correlation with GC methods (r-values between 88 and 98 percent). However, as the protocols supplied with the kits were somewhat confusing, the authors provided improved protocols for use by grainstore keepers, millers and malsters. It was also found that the operating ranges of the kits were not as wide as claimed by the manufacturer. In addition, some cross-reactivity of the antibodies was observed with structurally related compounds. Nonetheless, the study showed that ELISA tests could be used reliably to measure OP residues on stored grain.

The most common problems encountered in ELISAs are incomplete washing steps, pipetting problems, insufficient temperature control of reagents and plates, degradation of conjugate (may inhibit good binding to antibody) or loss of enzyme activity (Gee *et al.*, 1995). Problems with sample matrices are also common and can be detected by comparison of standards in an uncontaminated or 'blank' matrix. Recovery studies of standard additions may show up matrix problems. Alternatively, a dilution curve with sample matrix solution can be compared against a standard calibration curve. If the slopes are not parallel, a matrix effect is likely.

Finally, when purchasing an assay kit, the following criteria should be considered: price per assay, sensitivity, cross-reactivity, suitability for the chosen matrix, availability of published validation by independent workers and technical support (Gee *et al.*, 1995).

12. NEAR INFRARED REFLECTANCE SPECTROSCOPY (NIRS) FOR WHOLE SAMPLE ANALYSIS

Deaville and Flinn (2000) wrote a brief and clear introduction to the basic principles of NIRS. An accessible and more detailed review of applications and a description of the mathematical treatments of NIR spectra has been given by Shenk and Westerhaus (1995). Givens and Deaville (1999) have also reviewed NIRS in relation to feed analysis and animal nutrition.

The advantages of NIRS (Givens and Deaville, 1999) over traditional techniques are:

- Rapid, as minimal or no sample preparation necessary
- On-the-spot analysis of the whole sample, i.e. a non-destructive technique, allows simultaneous measurement of several parameters
- High precision
- High throughput makes NIR a cheap technique on a sample basis
- Environmentally friendly: no reagents, no chemical waste

The limitations of NIRS are:

- Suitable for major feed components, not for minor components
- Great care needed in developing calibrations as these are matrix specific
- Complexity in choice of data treatment is confusing for the novice

- Calibration procedures are time consuming and only worthwhile for subsequent analysis of large sample numbers
- High instrument costs

NIRS was developed in the 1950s and 1960s for quantitative analysis and applied to feed analysis in the 1970s (Norris *et al.*, 1976). By the late 1970s, NIRS was routinely used for protein measurements in grain. The NIRS region covers wavelengths between 730 and 2500 nm and is the infrared region that is particularly suited to quantitative analysis (Givens and Deaville, 1999). The main absorption bands of water are at 1940 and 1450 nm, of aliphatic C-H bonds at 2310, 1725, 1400 and 1210 nm, of O-H bonds around 2100 and 1600 nm and N-H bonds at 2180 and 2055 nm. Spectra are depicted in the form of the reciprocal log of reflectance (log1/R) and provide little direct information. Various components in the feed matrix produce a series of overlapping bands, which results in a smooth, rolling line. However, first- and second-order derivatives of the log1/R spectra can be used to resolve these overlapping bands.

NIRS data are generally subjected to a mathematical pretreatment to reduce interferences from light scatter (Barnes *et al.*, 1989). Then one of several different multivariate calibration methods is used to relate the spectral data from a sufficiently large and representative sample set to the primary, 'wet chemistry', data (Blanco *et al.*, 1997). Finally, calibrations are subjected to validation procedures with an independent set of samples. A simple monitoring procedure has been developed to minimize NIRS analysis errors (Shenk *et al.*, 1989).

Great care should be taken in developing NIR calibrations (references in Deaville and Flinn, 2000). Calibrations should be based on at least 50 samples, but often many more are required (>150). Givens and Deaville (1999) pointed out that 'NIRS is largely a secondary technique requiring calibration using samples of known composition determined by using standard methods (primary techniques)'. A problem can arise, when the primary methods do not define well the chemical constituent, e.g. drying at 100°C to determine moisture does not necessarily define water content or 6.25

x total N does not necessarily describe protein content adequately (Shenk and Westerhaus, 1995).

It has been stressed that consistent sample preparation is required, as variations in particle size, residual moisture content and packing density can adversely affect NIR spectra. However, new developments of NIR software, such as the noise repeatability file, have succeeded in reducing the sensitivity to residual sample moisture (Baker *et al.*, 1994; Shenk and Westerhaus, 1995). Alternatively, by using a coarse transport cell, a larger surface area of fresh grass silage can be screened, thus eliminating the need for dry silages (Park *et al.*, 1999a, b).

NIRS has been accepted as an official AOAC method for crude protein and ADF (AOAC 989.03) and for moisture (AOAC 991.01; Barton and Windham, 1998). It has also been used for determining starch and non-starch polysaccharides, fat and oil, metabolizable energy, insect or weed seed contamination in feed grains (Wrigley, 1999) and for the analysis of dried forages (Murray, 1993). It can be used to identify feeds and perform authenticity checks (De Boever *et al.*, 1993). In addition, heat damaged protein, fungal contamination and adulteration can be detected with modern pattern recognition software (Givens and Deaville, 1999).

It is also possible to transfer calibrations developed on an expensive scanning instrument to cheaper filter instruments in a network (Puigdomènech *et al.*, 1997). This ISI cloning software also allowed a successful transfer of NIR calibrations developed for fresh grass silages from a Foss to a Bran & Luebbe instrument (Park *et al.* 1999b).

Researchers are now aiming to predict directly the functional properties of feeds to animals, i.e. nutrient supply and production responses such as live weight gain, milk fat and protein or meat composition, rather than measuring feed components (Wrigley, 1999). Good predictions have been achieved for organic matter digestibility *in vivo* (Barber *et al.*, 1990); ME content (Givens *et al.*, 1992) and voluntary feed intake (see references in Deaville and

Flinn, 2000). It has also been possible to predict nutritionally relevant products, such as lactic acid, VFA's and cumulative gas volumes with calibrations based on 800 fresh silages (Deaville and Flinn, 2000). The ultimate aim is to formulate diets for optimum animal productivity, cost effectiveness and the least environmental effects (Givens and Deaville, 1999).

13. CONCLUSIONS

This review describes general procedures for feed analysis and includes recent developments towards faster techniques for fibre (FibreAnalyzer) and fat (Soxflo) analysis, new enzyme kits for rapid starch analysis, ELISAs for mycotoxins and other feed contaminants and progress in whole feed analysis by NIR spectroscopy.

It would appear that feedstuff analysis is difficult in comparison with other analyses (Midkiff, 1984; Key *et al.* 1997). The analysis of an individual, well-defined chemical compound tends to be easier to standardize between laboratories than some of the less well-defined, but nutritionally important parameters, e.g. oil, fibre, starch or digestibility of animal feeds. A better understanding of these parameters would help to improve their measurement.

Several inter-laboratory comparisons have demonstrated that variation tends to be unacceptably large, and that a coordinated effort has to be made to reach consensus procedures amongst feed analysis laboratories. Significant progress has been achieved through international efforts in producing CRMs for minerals, pesticides and pollutants in a range of environmental and food matrices. However, there are fewer CRMs for feeds, with still no CRM for NDF, and a dearth of proficiency schemes for animal feeds.

Plant breeders have access to several assays that determine levels of single, simple constituents responsible for beneficial or harmful effects (amino acids, n-3 fatty acids, minerals, mycotoxins, glucosinolates, alkaloids) (Wrigley 1999).

However, in cases of more complex mixtures of pro- or anti-nutrients, e.g. starches, tannins, lectins, structure-function relationships have not yet been elucidated and, therefore, no particular assay can be recommended for use in breeding programmes (Caygill and Mueller-Harvey, 1999). Similarly, tests for biologically active compounds, e.g. phytoestrogens or antioxidants, could be used more widely by plant breeders once the beneficial target compounds have been clearly identified.

Finally, the concept of good analytical practice, that ensures 'traceability of measurements' through use of validated methods and CRMs, has also been translated to 'traceability of feeds' by a feed manufacturer in the form of a 'Feed Passport'. This is likely to become increasingly important following recent experience with meat and bone meal in cattle feed and new soyabean varieties that led to the BSE crisis and GM disputes.

REFERENCES

AFRC. 1993. *Energy and protein requirements of ruminants.* CAB International, Wallingford, UK.

AOAC. 1988. Guidelines for collaborative study procedure to validate characteristics of a method of analysis. *Journal of the Association of Official Analytical Chemists,* 71: 161–171

AOAC. 1995. Animal Feed. Chapter 4. in *Official methods of analysi'.* 16th edition. AOAC International, Arlington, VI, USA. 30 pp.

Allen, M.S., Oba, M. & Choi, B.R. 1997. Nutritionist's perspective on corn hybrids for silage. In *Silage: Field to feedbank.* NRAES-99, Northeast Regional Agricultural Engineering Service, Ithaca, New York, pp. 25–36.

Åman, P. & Graham, H. 1990. Chemical evaluation of polysaccharides in animal feeds. Ch 9. *In* J. Wiseman & D.J.A. Cole, eds. *Feedstuff evaluation,* pp. 161–177. Butterworths, London.

Aufrère, J. & Michalet-Doreau, B. 1988. Comparison of methods for predicting digestibility of feeds. *Animal Feed Science and Technology,* 20: 203–218.

Bailey, S. & Henderson, K. 1990. Consequences of inter-laboratory variation in chemical analysis. Ch. 20. *In* J. Wiseman & D.J.A. Cole, eds. *Feedstuff evaluation*, pp. 353–363. Butterworths, London.

Baker, C.W., Givens, D.I. & Deaville, E.R. 1994. Prediction of organic matter digestibility *in vivo* of grass silage by near infrared reflectance spectroscopy: effect of calibration method, residual moisture and particle size. *Animal Feed Science and Technology*, 50: 17–26.

Barber, G.D., Givens, D.I., Kridis, M.S., Offer, N.W. & Murray, I. 1990. Prediction of the organic matter digestibiliy of grass silage. *Animal Feed Science and Technology*, 28: 115–128.

Barnes, R.J., Dhanoa, M.S. & Lister, S.J. 1989. Standard normal variate transformation and de-trending of near-infrared diffuse reflectance spectra. *Applied Spectroscopy*, 43: 772–777.

Barton, F.E., II & Windham, W.R. 1988. Determination of acid detergent fibre and crude protein in forages by near infrared reflectance spectroscopy: collaborative study. *Journal of the Association of Official Analytical Chemists*, 71: 1162–1167.

Beever, D.E., Cammell, S.B. & Edmonds, S. 1996. Inter-laboratory variation in foodstuff evaluation of two contrasting maize silage samples. *Animal Feed Science*, 62: 685–686.

Blanco, M., Coello, J. Iturriaga, S., Maspoch, S. & de la Pezuela, C. 1997. Calibration in near infrared diffuse reflectance spectroscopy. A comparative study of various methods. *Journal of Near Infrared Spectroscopy* 5, 67-75.

Brown, R.H. & Mueller-Harvey, I. 1999. Evaluation of the novel Soxflo technique for rapid extraction of crude fat in foods and animal feeds. *Journal of AOAC International* 82, 1369-1374.

Butter, N.L., Dawson, J.M. & Buttery, P.J. 1999. Effects of dietary tannins on ruminants. Chapter 5. *In* J.C. Caygill & I. Mueller-Harvey, eds. *Secondary plant products – antinutritional and beneficial actions in animal feeding*, pp. 51–70. Nottingham University Press, Nottingham, UK.

Caygill, J.C. & Mueller-Harvey, I. 1999. *Secondary Plant Products - Antinutritional and beneficial actions in animal feeding*. Nottingham University Press, Nottingham, UK, 129 pp.

Chai, W.H. & Uden, P. 1998. An alternative oven method combined with different detergent strengths in the analysis of neutral detergent fibre. *Animal Feed Science and Technology*, 74: 281–288.

Cherney, D.J.R. 2000. Characterization of forages by chemical analysis. Ch 14. *In* D.I. Givens, E. Owen, R.F.E. Axford & H.M. Ohmed, eds. *Forage evaluation in ruminant nutrition*, pp. 281–300. CABI Publishing, Wallingford, UK.

Cherney, D.R.J., Patterson, J.A. & Cherney, J.H. 1989. Use of 2-ethoxyethanol and a-amylase in the neutral detergent fiber method of feed analysis. *Journal of Dairy Science*, 72: 3079–3084.

Cochrane, W.P. 1991. Testing of food and agricultural products by immunoassay. Chapter 4. *ACS Symposium Series*, 451: 40–48.

Crosby, N. 1995. Animal Feeds. *In* A. Townshend, ed. *Encyclopedia of analytical science*, Vol. 1, pp. 120–136.Academic Press, London.

Dawra, R.K., Makkar, H.P.S. & Singh, B. 1988. Protein-binding capacity of microquantities of tannins. *Analytical Biochemistry*, 170: 50–53.

Deaville, E.R. & Flinn, P.C. 2000. Near-infrared (NIR) spectroscopy: an alternative approach for the estimation of forage quality and voluntary intake. Chapter 15. *In* D.I.Givens, E. Owen, R.F.E. Axford & H.M. Ohmed, eds.*Forage evaluation in ruminant nutrition*, pp. 301–320. CABI Publishing, Wallingford, UK.

De Boever, J.L., Cottyn, B.G., Vanacker, J.M. & Boucque, C.V. 1994. An improved enzymatic method by adding gammanase to determine digestibility and predict energy value of compound feeds and raw materials for cattle. *Animal Feed Science and Technology*, 47: 1–18.

Edmunds, B.K. 1990. Chemical analysis of lipid fractions. Ch 11. *In* J. Wiseman & D.J.A. Cole, eds. *Feedstuff evaluation*, pp. 197–213. Butterworths, London.

EURACHEM/WELAC. 1998. Guidance Document. Accreditation for Chemical Laboratories. Guidance on the interpretation of the EN 45000 series of standards and ISO/IEC guide 25.

Fahey, G.C., Collins, M., Mertens, D.R. & Moser, L.E. 1994. *Forage quality, evaluation, and utilization*. American Society of Agronomy, Madison, WI, USA.

Faichney, G.J. & White, W.T. 1983. *Methods for the Analysis of Feeds Eaten by Ruminants*. CSIRO, Melbourne, Australia, 36 pp.

Feeding Stuffs. 1988. *The Feeding Stuffs (Sampling and Analysis) Regulations*, 1988 (as amended). Statutory Instrument No 1144. HMSO, London, UK.

Folin, O. & Ciocalteau, V. 1927. On tyrosine and tryptophane determination in proteins. *Journal of Biological Chemistry*, 73: 424–427.

Gangaiya, P. & Morrison, J. 1992. Problems of establishing quality assurance in developing country laboratories. LGC plc, Teddington, UK. *VAM Bulletin* 7, 13–14.

Gee, S.J., Hammock, B.D. & Skerritt, J.H. 1995. Diagnostics for plant agrochemicals - a meeting of chemistry and immunoassay. Chapter 11. *In* J.H. Skerritt & R. Appels, eds. *New diagnostics in crop sciences*, pp. 243–276. CAB International, Wallingford, UK.

Giner-Chavez, B.I., Van Soest, P.J., Robertson, J.B., Lascano, C. Reed, J.D. & Pell, A.N. 1997. A method for isolating condensed tannins from crude plant extracts with trivalent ytterbium *Journal of the Science of Food and Agriculture*, 74: 359–368.

Givens, D.I. & Deaville, E.R. 1999. The current and future role of near infrared reflectance spectroscopy in animal nutrition: a review. *Australian Journal of Agricultural Research*, 50: 1131–1145.

Givens, D.I., Baker, C.W. & Zamime, B. 1992. Regions of normalised near infrared reflectance difference spectra related to the rumen digestion of straws. *Animal Feed Science and Technology*, 36: 1–12.

Givens, D.I., Owen, E., Axford, R.F.E. & Ohmed, H.M. 2000a. *Forage evaluation in ruminant nutrition*. CABI Publishing, Wallingford, UK.

Givens, D.I., Owen, E. & Adesogan, A.T. 2000b. Current procedures, future requirements and the need for standardization. Ch. 21. *In* D.I. Givens, E. Owen, R.F.E. Axford & H.M. Ohmed, eds. *Forage evaluation in ruminant nutrition*, pp. 449–474. CABI Publishing, Wallingford, UK.

Goering, H.K. & Van Soest, P.J. 1970. Forage Fiber Analyses (Apparatus, Reagents, Procedures, and Some Applications). *Agricultural Handbook* 379, ARS, USDA, Washington, DC. 20 pp.

Graham, H.D. 1992. Stabilization of the Prussian blue color in the determination of polyphenols. *Journal of Agricultural and Food Chemistry,* 40: 801–805.

Guyot, S., Doco, T., Souquet, J-M., Moutounet, M. & Driliau, J-F. 1997. Characterization of highly polymerized procyanidins in cider apple (*Malus sylvestris* var. Kermerrien) skin and pulp. *Phytochemistry,* 44: 351–357.

Hagerman A.E. Web site: http://miavx1.muohio.edu/~hagermae/

Hagerman, A.E. & Butler, L.G. 1980. Determination of protein in tannin-protein precipitates. *Journal of Agricultural and Food Chemistry,* 28: 944–947.

Hagerman, A.E., Robbins, C.T., Weerasuriya, Y., Wilson, T.C. & McArthur, C. 1992. Tannin chemistry in relation to digestion. *Journal of Range Management,* 45: 57–62.

Hagerman, A.E., Zhao, Y. & Johnson, S. 1997. Methods for determination of condensed and hydrolyzable tannins. Chapter12. *In* F. Shahidi, ed. *Antinutrients and phytochemicals in food*, pp. 209–222. American Chemical Society, ACS Symposium Series 662.

Hedqvist, H., Mueller-Harvey, I., Reed, J.D., Krueger, C.G. & Murphy, M. 2000. Characterisation of tannins and *in vitro* protein digestibility of several *Lotus corniculatus* varieties. *Animal Feed Science and Technology,* 87: 41–56.

Hintz, R.W. & Mertens, D.R. 1996. Effects of sodium sulfite on recovery and composition of detergent fiber and lignin. *Journal of AOAC International,* 79: 16–22.

Horwitz, W. 1993. International coordination and validation of analytical methods. *Food Additives and Contaminants,* 10: 61–69.

Jackson, F.S., McNabb, W.C., Barry, T.N., Food, Y.L. & Peters, J.S. 1996. The condensed tannin content of a range of subtropical and temperate forages and the reactivity of condensed tannin with ribulose-1,5-bis-phosphate carboxylase (Rubisco) protein. *Journal of the Science of Food and Agriculture,* 72: 483–492.

Jones, R.J. & Palmer, B. 2000. *In vitro* digestion studies using 14C-labelled polyethylene glycol (PEG) 4000: comparison of six tanniniferous shrub legumes and the grass *Panicum maximum. Animal Feed Science and Technology* 85, 215-221.

Jones, W.R., Broadhurst, R.B. & Lyttleton, J.W. 1976. The condensed tannins of pasture legume species. *Phytochemistry,* 15: 1407–1409.

Kawamura, O., Wang, D.S., Liang, Y.X., Hasegawa, A., Saga, C., Visconti, A. & Ueno, Y. 1994. Further survey of aflatoxin M1 in milk powders by ELISA. *Food and Agricultural Immunology,* 6: 465–467.

Key, P.E., Patey, A.L., Rowling, S., Wilbourn, A. & Worner F.M. 1997. International proficiency testing of analytical laboratories for foods and feeds from 1990 to 1996: the experiences of the United Kingdom food analysis performance assessment scheme. *Journal of AOAC International,* 80: 895–899.

Krueger, C.G., Dopke, N,C., Treichel, P.M., Folts, J. & Reed, J.D. 2000a. Matrix-assisted laser desorption/ionization time-of-flight mass spectrometry of polygalloyl polyflavan-3-ols in grape seed extract. *Journal of Agricultural and Food Chemistry,* 48: 1663–1667.

Krueger, C.G., Porter, M.L., Wiebe, D.A., Cunningham, D.G. & Reed, J.D. 2000b. Potential of cranberry flavonoids in the prevention of copper-induced LDL oxidation. Freising-Weihenstephan (Germany), September 11-15, 2000. *Polyphenols Communications,* 2: 447–448.

Lakin, A.L. 1978. Determination of nitrogen and estimation of protein in foods. *In* R.D. King, ed. *Developments in Food Analysis Techniques.* Vol 1, pp. 43–74. Applied Science Publishers, London.

Lanari, D., Pinosa, M., Tibaldi, E. & D'Agaro, E. 1991. Risultati di due ring test condotti sulla composizione chimica di alcuni alimenti e sulla digeribilità *in vivo. Zootecnica e Nutrizione Animale,* 17: 285–295.

Li, Y-G, Tanner, G. & Larkin, P. 1996. The DMACA-HCl protocol and the threshold proanthocyanidin content for bloat safety in forage legumes. *Journal of the Science of Food and Agriculture,* 70: 89–101.

Lowry, J.B., McSweeney, C.S. & Palmer, B. 1996. Changing perceptions of the effect of plant phenolics on nutrient supply in the ruminant. *Australian Journal of Agricultural Research,* 47: 829–842.

Madsen, J. & Hvelplund, T. 1994. Prediction of *in situ* protein degradability in the rumen. Results of a European ringtest. *Livestock Production Science,* 39: 201–212.

Madsen, J., Hvelplund, T. & Weisbjerg, M.R. 1997. Appropriate methods for the evaluation of tropical feeds for ruminants. *Animal Feed Science Technology*, 69: 53–66.

MAFF. 1986. *The analysis of agricultural materials*. Reference Book 427. 3rd edition. HMSO. 248pp.

Maier, E.A., Griepink, B., Quevauviller, P., De Angelis, L. & Muntau, H. 1990. Certified reference materials - hay powder (CRM 129) and rye grass (CRM 281) - for the quality control of analysis of animal feed. *Mikrochimica Acta* (Wien) III, 87–99.

Marais, J.P.J., Mueller-Harvey, I., Brandt, E.V. & Ferreira, D. 2000. Polyphenols, condensed tannins and other natural products in *Onobrychis viciifolia* (Sainfoin). *Journal of Agricultural and Food Chemistry*, 48: 3440–3447.

Makkar, H.P.S., Dawra, R.K. & Singh, B. 1987. Prtoein precipitation assay for quantitation of tannins: determination of protein in tannin-protein complex. *Analytical Biochemistry*, 166: 435–439.

Makkar, H.P.S., Blümmel, M., Borowy, N.K. & Becker, K. 1993. Gravimetric determination of tannins and their correlations with chemical and protein precipitation methods. *Journal of the Science of Food and Agriculture*, 61: 161–165.

Matthews, M.A., Haverley, M. & Betteridge, J.W. 1996. Immunoassays for the detection of organophosphorus pesticides on stored grain: assessment of three commercially available kits and reocmmendations for laboratory and field use. Home-Grown Cereals Authority, London, UK. Project Report No. 122. 46 pp.

Mauricio, R.M., Mould, F.L., Dhanoa, M.S., Owen, E., Channa, K.S. & Theodorou, M.K. 1999. A semi-automated *in vitro* gas production technique for ruminant feedstuff evaluation. *Animal Feed Science and Technology*, 79: 321–330.

McQueen, R.E. & Nicholson, R. 1979. Modification of the neutral detergent fibre procedure for cereals and vegetables by using a-amylase. *Journal of the Association of Official Analytical Chemists*, 62: 676–680.

Midkiff, V. 1984. A century of analytical excellence. The history of feed analysis, as chronicled in the development of AOAC official methods,

1884-1984. *Journal of the Association of Official Analytical Chemists,* 67: 851–860.

Minson, D.J. 1990. *Forage in ruminant nutrition.* Academic Press, San Diego, CA, USA.

Moore, K.I. & Hatfield, R.D. 1994. Carbohydrates and forage quality. Chapter 5. *In* G.C. Fahey, M. Collins, D.R. Mertens & L.E. Moser, eds. *Forage quality, evaluation, and utilization,* pp. 229-280. American Society of Agronomy, Madison, WI, USA.

Morgan, M.R.A. 1995. Monitoring safety of plant foods: immunodiagnostics for mycotoxins and other bioactive compounds. Chapter 10. *In* J.H. Skerritt & R. Appels, eds. *New diagnostics in crop sciences,* pp. 215–242. CAB International, Wallingford, UK.

Mueller-Harvey, I. 1999. Tannins: their nature and biological significance. Chapter 3. *In* J.C. Caygill & I. Mueller-Harvey, eds. *Secondary plant products – antinutritional and beneficial actions in animal feeding,* pp. 17–39. Nottingham University Press, Nottingham, UK.

Mueller-Harvey, I. 2001. Analysis of hydrolysable tannins - a review. *Animal Feed Science Technology* (in press).

Mueller-Harvey, I. & McAllan, A.B. 1992. Tannins - their biochemistry and nutritional properties. *In* I.M. Morrison, ed. *Advances in plant cell biochemistry and biotechnology,* Vol 1, 151–217. JAI Press, London.

Mueller-Harvey, I., Reed, J.D. & Hartley, R.D. 1987. Characterisation of phenolic compounds, including flavonoids and tannins, of ten Ethiopian browse species by high performance liquid chromatography. *Journal of the Science of Food and Agriculture,* 39: 1–14.

Murray, I. 1993. Forage analysis by near infrared reflectance spectroscopy. *In* A. Davies, R.D. Baker, S.A. Grant & A.S. Laidlaw, eds. *Sward management handbook,* pp. 285–312. British Grassland Society, Reading, UK.

Norris, K.H., Barnes, R.F., Moore, J.E. & Shenk, J.S. 1976. Predicting forage quality by infrared reflectance spectroscopy. *Journal of Animal Science,* 43: 889–897.

Okuda, T., Yoshida, T. & Hatano, T. 1989. New methods of analyzing tannins. *Journal of Natural Products,* 52: 1–31.

Park, R.S., Agnew, R.E. & Barnes, R.J. 1999a. The development of near infrared reflectance spectroscopy calibrations for undried grass silage and their transfer to another instrument using multiple and single sample standardisation. *Journal of Near Infrared Spectroscopy,* 7: 117–131.

Park, R.S., Agnew, R.E., Gordon, F.J. & Barnes, R.J. 1999b. The development and transfer of undried grass silage calibrations between near infrared reflectance spectroscopy instruments. *Animal Feed Science and Technology,* 78: 325–340.

Patey, A. 1996. Quality in quantities. *Chemistry in Britain,* 32: 35–37.

Pell, A.N & Schofield, P. 1993. Computerized monitoring of gas production to measure forage digestion *in vitro. Journal of Dairy Science,* 76: 1063–1073.

Petterson, D.S., Harris, D.J., Rayner, C.J., Blakeney, A.B. & Choct, M. 1999. Methods for the analysis of premium livestock grains. *Australian Journal of Agricultural Research,* 50: 775–787.

Porter, L.J., Hristich, L.N. & Chan, B.G. 1986. The conversion of procyanidins and prodelphinidins to cyanidin and delphinidin. *Phytochemistry,* 25: 223–230.

Puigdomènech, A., Tauler, R., Casassas, E. & Aragay, M. 1997. Modelling near infrared instrument differences by chemometric methods: testing for near infrared forage analysis. *Analytica Chimica Acta,* 355: 181–193.

Reed, J.D. 1986. Relationships among soluble phenolics, insoluble proanthocyanidins and fiber in East African browse species. *Journal of Range Management,* 39: 5–7.

Reed, J.D., Horvath, P.J., Allen, M.S. & Van Soest, P.J. 1985. Gravimetric determination of soluble phenolics including tannins from leaves by precipitation with trivalent ytterbium. *Journal of the Science of Food and Agriculture,* 36: 255–261.

Reed, J.D., Krueger, C., Rodriguez, G. & Hanson, J. 2000. Secondary plant compounds and forage evaluation. Chapter 20. *In* D.I.Givens, E. Owen, R.F.E. Axford & H.M. Ohmed, eds. *Forage evaluation in ruminant nutrition,* pp.433–448. CABI Publishing, Wallingford, UK.

Reynolds, C.K., Sutton, J.D. & Beever, D.E. 1997. Effects of feeding starch to dairy cattle on nutrient avialability and production. *In:* P.C. Garnsworthy & J. Wiseman, eds. *Recent advances in animal nutrition*, pp 105–134. Butterworths, London.

Schingoethe, D.J. 1991. Pyproduct feeds: feed analysis and interpretation. *Veterinary Clinics of North America: Food Animal Practice*, 7: 577–584.

Schneider, E., Dietrich, R., Martlbauer, E., Usleber, E. & Terplan, G. 1991. Detection of aflatoxins, trichothecenes, ochratoxin A and zearalenone by test strip immunoassay: a rapid method for screening cereals for mycotoxins. *Food and Agricultural Immunology*, 3: 185–193.

Schofield, P., Mbugua, D.M. & Pell, A.N. 2001. Analysis of condensed tannins: a review. *Anim. Feed Sci. Technol.* (in press).

Shenk, J.S. & Westerhaus, M.O. 1995. The application of near infrared reflectance spectroscopy (NIRS) to forage analysis. Chapter 10. *In* G.C. Fahey, M. Collins, D.R. Mertens & L.E. Moser, eds. *Forage quality, evaluation, and utilization*, pp. 406–449. American Society of Agronomy, Madison, WI, USA.

Shenk, J.S., Westerhaus, M.O. & Abrams, S.A. 1989. Protocol for NIRS calibration: monitoring analysis results and recalibration. *In* G.C. Marten, J.S. Shenk & F.E. Barton, II, eds. *Near infrared reflectance spectroscopy (NIRS): analysis of forage quality*, pp. 104–110. Agric. Handbook No. 643. USDA-ARS. U.S. Government Printing Office, Washington, DC.

Silanikove, N, Shinder, D, Gilboa, N, Eyal, M. & Nitsan, Z. 1996. Polyethylene glycol binding to plant samples as an assay for the biological effects of tannins: predicting the negative effects of tannins in Mediterranean browse on rumen degradation. *Journal of Agricultural and Food Chemistry*, 44: 3230–3234.

Simonne, A.H., Simonne, E.H., Eitenmiller, R.R., Mills, H.A. & Cresman III, C.P. 1997. Could the Dumas method replace the Kjeldahl digestion for nitrogen and crude protein determinations in foods? *Journal of the Science of Food and Agriculture*, 73: 39–45.

Skerritt, J.H. & Appels, R. 1995. An overview of the development and application of diagnostic methods in crop sciences. Chapter 1. *In* J.H. Skerritt and R. Appels, eds. *New diagnostics in crop sciences*, pp. 1–32. CAB International, Wallingford, UK.

Stewart, J.L., Mould, F. & Mueller-Harvey, I. 2000. The effect of drying treatment on the fodder quality and tannin content of two provenances of *Calliandra calothyrsus* Meissner. *Journal of the Science of Food and Agriculture*, 80: 1461–1468

Strong, F.C. III & Duarte, A.M.A. 1992. A room-temperature, rapid method for the determination of protein in wheat and other grains by the biuret reaction. *Cereal Chemistry*, 69: 659–664.

Tanaka, T., Sueyasu, T., Nonaka, G-I. & Nishioka, I. 1984. Tannins and related compounds. XXI. Isolation and characterization of galloyl and p-hydroxybenzoyl esters of benzophenone and xanthone C-glucosides from *Mangifera indica* L. *Chem. Pharm. Bull.*, 32: 2676–2686.

Terrill, T.H., Windham, W.R., Evans, J.J. & Hoveland, C.S. 1990. Condensed tannin concentration in *Sericea lespedeza* as influenced by preservation method, *Crop Science*, 301: 219–224.

Thiex, N., Smallidge, R. & Beine, R. 1996. Sources of error in vitamin A analysis. *Journal of AOAC International*, 79: 1269–1275.

Trucksess, M.W., Stack, M.E., Nesheim, S., Park, D.L. & Pohland, A.E. 1989. Enzyme-linked immunosorbent assay of aflatoxins B1, B2, and G1 in corn, cottonseed, peanuts, peanut butter, and poultry feed: collaborative study. *Journal of the Association of Official Analytical Chemists*, 72: 957–962.

Trucksess, M.W., Young, K., Donahue, K.F., Morris, D.K. & Lewis, E. 1990. Comparison of two immunochemical methods with thin-layer chromatographic methods for determination of aflatoxins. *Journal of the Association of Official Analytical Chemists*, 73: 425–428.

Undersander, D. Mertens, D.R. & Thiex, N. 1993. Forage analyses procedures. National Forage Testing Association, Omaha, NE. [see also: http://www.uwex.edu/ces/forage/nfta/ndfiber.htm]

Von Holst, C., Pallaroni, L, Björklund, E., Unglaub, W. & Anklam, E. 2000. Control of appropriate heat treatment of animal waste using the

ELISA technique: validation and calibration of an analytical method. *In: Proceedings of the 6th International Feed Production Conference. Food Safety: current situation and perspectives in the European Community*. 27-28th November 2000. Pisa, Italy. pp. 96–105.

Van Soest, P.J., Robertson, J.B. & Lewis, B.A. 1991. Methods for dietary fiber, neutral detergent fiber, and nonstarch polysaccharides in relation to animal nutrition. *Journal of Dairy Science*, 74: 3583-3597.

Walker and Brookman. 1998. *Guidelines for the in-house production of reference materials*. LGC/VAM/1998/040. LGC plc, Teddington, UK. 32 pp.

Waterman, P.G. & Mole, S. 1994. *Analysis of phenolic plant metabolites*. Blackwell, Oxford.

Watson, C.A. 1994. *Official and standardized methods of analysis*. 3rd edition. Royal Society of Chemistry, Cambridge, UK. 778 pp.

Weigert P., Gilbert, J., Patey, A.L., Key, P.E., Wood, R. & Barylko-Pikielna, N. 1997. Analytical quality assurance for the WHO GEMS/ Food EURO programme - results of 1993/94 laboratory proficiency testing. *Food Additives and Contaminants*, 14: 399–410.

Williams, A. 1996. Measurement uncertainty in analytical chemistry. *Accreditation and Quality Assurance*, 1: 14–17:

Willis, R.B. & Allen, P.R. 1998. Improved method for measuring hydrolyzable tannins using potassium iodate. *Analyst*, 123: 435–439.

Wilson, T.C. & Hagerman, A.E. 1990. Quantitative determination of ellagic acid. *Journal of Agricultural and Food Chemistry*, 38: 1678–1683.

Wiseman, J. & Cole, D.J.A. 1990. *Feedstuff evaluation*. Butterworths, London.

Wrigley, C.W. 1999. Potential methodologies and strategies for the rapid assessment of feed-grain quality. *Australian Journal of Agricultural Research*, 50: 789–805.

Yanagida, A., Kanda, T., Shoji, T., Ohnishi-Kameyama, M. & Nagata, T. 1999. Fractionation of apple procyanidins by size-exclusion chromatography. *Journal of Chromatography*, 855: 181–190.

Zeeman & Bonn. 1995. Carbohydrates. *In* A. Townshend, ed), *Encyclopedia of analytical science*, Vol. 1. pp 451 – 525. Academic Press, London.

Appendix 1

1. USEFUL WEBSITES FOR FURTHER INFORMATION (TEACHING AND TRAINING)

http://www.lgc.co.uk/best/terp/terpoindex.htm
http://www.agric.rdg.ac.uk/Facilities/FAL/FAL_frameset.htm
'Accreditation and Quality Assurance' - a journal for quality, comparability and reliability in chemical measurement.

2. WEB SITES FOR ANALYTICAL METHODS

Methods for forage testing:
http://www.dfrc.wisc.edu/foragetesting_nfta.html
Methods for tannin analysis: http://miavx1.muohio.edu/~hagermae/

2.3 Addresses and Web sites for Certified Reference Materials

National Institute of Standards & Technology
Standard Reference Materials Programme
Building 202, Room 204
Gaithersburg, MD 20878-9950, USA
Fax: +1 301 948 3730

Community Bureau of Reference
Commission of the European Communities
Rue de la Loi 200
B-1049 Brussels, Belgium
Fax: +32 2 295 8072

The Office of Reference Materials
Laboratory of the Government Chemist
Queens Road
Teddington

Middlesex TW11 0LY, U.K.
e-mail: orm@lgc.co.uk
http://www.lgc.co.uk

Promochem Ltd
6 South Mundells
Haslemere Industrial Estate
Welwyn Garden City
Herts AL7 1EP, U.K.
Fax: +44 1707 396 677

Breitlaender Eichproben und Labormaterial
GmbH
Postfach 8046
4700 Hamm 3, Germany
Fax: +49 2381 403 189

National Institute for Environmental Studies
Japan Environment Agency
P O 16-2 Onogawa
Tsukuba
Ibaraki 305, Japan
Fax: (0298) 51 6111

2.4 Addresses for RING TESTS or Proficiency Schemes

National Forage Testing Association, USA:
http://www.dfrc.wisc.edu/foragetesting_
nfta.html

Dr V Houba
Department of Soil Science and Plant
Nutrition
Wageningen Agricultural University
P O Box 8005

6700 EC Wageningen/ The Netherlands
Fax: +31 8370 83766

Dr A. Eijgenraam
Bedrijfslaboratorium voor Grond- en
Gewasonderzoek
Postbus 115
6860 Oosterbeek, The Netherlands
Fax: 0031 26 333 7831
(Please note: two feed samples per year for non-commercial laboratories only)

ALASA
P O Box 14105
Verwoerdburg 0140
Republic of South Africa
Fax: +27 12 664 1431

Dr A.L. Patey
FAPAS Secretariat
Ministry of Agriculture, Fisheries and Food
Food Science Laboratory
Norwich Research Park
Colney
Norfolk NR4 7UQ, U.K.
Fax: +44 1603 501 123

2.5 Web sites for Accreditation

http://www.european-accreditation.org
http://www.aoac.org/techprog/Intro98.htm
http://www.ukas.com/
http://www.info.gov.hk/itc/eng/quality/hkas/hkas.shtml

Variability in feed composition and its impact on animal production

G. Gizzi and D.I. Givens
ADAS Nutritional Sciences Research Unit
Stratford-upon-Avon
United Kingdom

Knowledge of feedstuff composition is vital for the nutritionist, in order to meet precisely the nutrient requirements of livestock; the feed manufacturer, in order to produce balanced compound feeds; the farmer, in order to plan forage production; and the policy-maker, in order to direct strategies that guarantee a competitive, sustainable, environmentally friendly and food-safe agriculture. Feed tables and databases of feed composition and the nutritional value of feedstuffs can provide this information. However, such collections of data are valuable only as long as they provide reliable and up-to-date information. Data on feedstuff characteristics may be affected by different types of variability. The use of imprecise feedstuffs data, characterized by low intrinsic and high extrinsic variability, is deleterious to animal production from the nutritional, economical and environmental points of view.

1. INTRODUCTION

In order to meet the nutritional requirements of livestock a precise knowledge of feedstuff composition is necessary. Such information is particularly vital when trying to achieve the high levels of production required in today's competitive markets (where there is growing interest in quality, efficiency and the environment) through the preparation of balanced diets for the animals. Knowledge is also essential in order to mix the right proportions of various ingredients for the manufacture of valuable compound feeds. Ultimately, knowing the chemical characteristics and the nutritional value of raw materials is fundamental in the planning of forage production on the farm so that crop yields can be balanced with the animals' requirements. The general aspects of this have been reviewed by Topps (1989).

From a political point of view, data on feedstuff composition and nutritional value provide essential information to policy-makers that allows them to develop competitive and sustainable agriculture.

Feedstuffs are usually described by their chemical composition as well as by their nutritional value. Information on the composition of a feedstuff is acquired through chemical analysis, and most commonly includes such parameters as dry matter, protein, fiber fractions, organic matter and fat contents. The nutritional value of a feedstuff is assessed from experiments with animals (*in vivo* digestibility/*in situ* degradability, etc.) and provides information on how feedstuffs are digested and metabolized by the animal, mainly through interpreting differences between the input and the output of a series of components. Among these, the digestible/degradable organic matter and the digestible or metabolizable energy values are the ones most frequently represented in databases and used to predict energy supply from feeds and diets.

To provide both chemical and nutritional information, feedstuffs have been analysed in laboratories and animal-based studies and the resulting data collected in tables of feed composition for more than 200 years. The first recorded publication was that of Thaer (1809). More recently, collections of data have been organized into computer programs, creating databases of feedstuffs composition. The first database available on a computer was produced in 1963 at Utah State University in the United States (Harris, Asplund and Crampton, 1968). Today, modern databases are powerful tools with which information can be rapidly retrieved, sorted, updated and printed.

2. SOURCES OF DATA

Information on feedstuffs composition and nutritional value plays a fundamental role at the farm, feed manufacturer and government levels, so the data inserted into databases need to be reliable and of high quality. The quality of the data depends, in the first place, on the origin of the information, which may have been extracted from the literature, collected from laboratories, obtained from other databases or generated specifically from animal studies and analyses of feeds.

Data assembled from the literature

Feedstuff information in the literature usually concentrates more on the data value obtained and does not give a full description of the feedstuffs concerned (Leche, 1983). When feedstuffs are insufficiently or inappropriately described there is a risk of confusing their real origin. A typical example of this occurs when publications generally refer to the product "distillers' dark grains", which is a by-product of the whisky distilling industry that, depending on the type of distilling process, may be derived from barley (in the malt distilling process) or wheat and/or maize (in the grain distilling process). In this case a true and existing genetic variability is ignored. It is evident that the compositions of the two products will differ, so the provision of additional information on the origin of the feedstuff is essential; it is not enough merely to refer to it as "distillers' dark grains". In the absence of additional information, when trying to obtain information on the protein content specifically of barley distillers' dark grains, a value obtained from the average protein content of samples of both barley dark grains and wheat/maize dark grains would be retrieved from the database, and this value would be meaningless. Data obtained from the literature can therefore often be inappropriate, particularly when they use averaged data which may incorrectly merge different feedstuffs. Data from the literature are recommended only when no other sources are available.

Data collected from laboratories

When compiling databases of feedstuffs information, the data used have usually been produced in a number of different laboratories. When assembling collections of data, it is necessary to have specific information regarding the analytical procedures used to obtain the data, in order to guarantee that those used to create a database have been obtained through standardized methods, are expressed in standardized units and can, therefore, be compared.

The standardization issue is particularly important for data on the nutritional value of feedstuffs, obtained through *in vivo* digestibility or *in situ* degradability studies. In this case, together with the strictly analytical issue, the laboratories and research institutes involved should have standardized procedures for conducting animal studies. These relate to the numbers of animals and the species used (sheep, cattle, pigs, etc.) and to the feeding and collection protocols employed. However, when conducting *in vivo/in situ* studies, results will always be substantially influenced by variability among individual animals, no matter what level of standardization is achieved.

The standardized analytical procedures used to collect data from laboratories need to be monitored continuously, and the data they produce checked for accuracy and relevance. For this purpose, ring tests among participating laboratories should be carried out in order to evaluate the precision of standard or newly developed techniques (Barber, 1983; Fisher, 1983).

Merging of new and old data

It is common practice to merge new databases with old ones. In these cases it is important that the analytical procedures used to obtain all the data are understood in order to generate information that is expressed on a common basis and is, therefore, comparable. The evolution of analytical methodology is a substantial issue. Old data may have been obtained with obsolete methods or with surviving analytical procedures that, in the past, produced data that were less accurate than those produced today. Thus, the merging of old data with new may be inappropriate and may reduce the overall quality of the information.

Data produced specifically for use in databases

The data generated specifically for use in databases are likely to be those of the highest quality as all the necessary information will be included and the methods standardized and checked. The chemical composition of feeds can be analysed using the most up-to-date and standard methods and animal studies can be designed appropriately, by deciding which animal species to use and in which physiological state the animals studied should be. Both feeding and collection protocols can also be set with the aim of meeting the specific requirements of the database.

Ideally, an entire database would be produced by analysing specifically the composition of feedstuffs and studying their nutritional value through animal studies. In this way the variability arising from extrinsic factors such as analysis methods and animal protocols would be reduced to the minimum. On the other hand, when using data from a single source it is critical that the techniques and procedures used have been certified for precision and are representative. However, if it is to contain representative information, the database has to be established with an appropriate number of samples. The costs involved in analyzing feedstuffs and, particularly, performing animal measurements are so prohibitive that it is rare for feedstuff information to be produced specifically for the purpose of assembling a database. Moreover, much time is required to obtain the data, particularly those derived from animal studies. Therefore, the data are usually extracted from the different sources listed previously. Ultimately, it is the responsibility of the database manager to judge which information to include in order to generate a meaningful database.

3. SOURCES OF DATA VARIABILITY

Information on feedstuffs may be associated with different sources and types of variability. Variability can be "intrinsic" and caused by real differences among feedstuffs, such as the genetic origin of the feed or the processes that have been applied to it. Variability can also be "extrinsic" that is caused by differences in sampling and analysis procedures.

Intrinsic variability of data

Each feed has particular chemical and nutritional characteristics which distinguish it from other feeds. However, the same feed can also be derived from different cultivars or influenced by its geographical origin. Moreover, an identical raw material may have been processed in a variety of ways to produce very different products. The wide range of diverse wheat-related materials provides a clear illustration of the need to distinguish among different feedstuffs derived from the same raw material. Terms for wheat-based materials include wheat, durum wheat, winter wheat, spring wheat, processed wheat meal and wheat distillers' dark grains. All of these products and by-products are wheat-based, but they vary substantially in terms of both chemical composition and nutritional attributes and it is therefore expected that they will have different effects when fed to animals.

When assembling a database it is necessary to make sure that the real variability arising from the intrinsic characteristics of the feedstuffs is correctly recognized in order not to affect the quality of the data. Information from feedstuffs databases is most commonly retrieved in the form of averages (or "consolidated data"), so the feeds have to be merged judiciously in order to obtain meaningful information. To achieve reliability, a precise description and the correct naming of feedstuffs are essential, and for this purpose all available information on the type, origin, processing, etc. of the feedstuffs is needed.

One important attribute that is often omitted is information on the forage's stage of maturity. As shown in the example illustrated in the Figure, the content of both neutral detergent fiber (NDF) and starch relate strongly to the stage of maturity of the forage maize, and this is clearly reflected in the dry matter (DM) content of the plant, which increases with maturation. In this case, therefore, it would be incorrect to describe material harvested at different stages simply as forage maize; the different maturity stages should be indicated and, on the basis of these, the different types of product classified.

With respect to the processes applied to the feeds, when introducing data on, for example, rice bran meal, it is necessary to know how the meal was obtained and which processes were involved in its production. In the example shown in Table 1, at least two different types of products (expelled and extracted rice bran meal)

FIGURE 1

Relationship between maturity stage and neutral detergent fibre (NDF) and starch contents of maize (from Givens *et al.*, 1995)

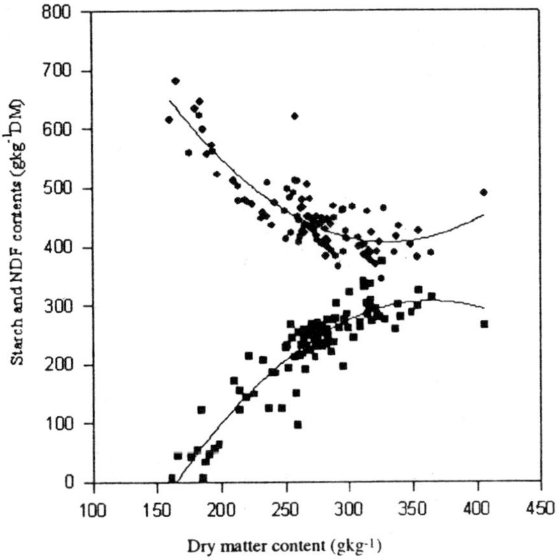

need to be differentiated because they are characterized by different chemical compositions and, consequently, by different nutritional values.

As shown in Table 1, the process that produces the extracted rice bran meal leads to a material with a much lower fat content compared with the expelled one (7.3 g/kg compared with 90.0 g/kg DM), and this produces a direct effect on the metabolizable energy value. It is apparent that if, instead of specifying the type of process used, the values of the two types of rice bran meal had been averaged, the resulting values would have been meaningless.

Another, simpler, example is related to the on-farm processes applied to forages. Hay is obtained by drying the forage, but the drying process may be achieved by "sun curing" or "barn curing" the

TABLE 1
Extracted and expressed rice bran meal: analytical and nutritional differences

	Extracted meal	Expressed meal
Analytical database parameters		
Oven DM (fresh basis)	896	902
Crude protein (g/kg DM)	54	128
NDF1 (g/kg DM)	450	370
Ether extract (g/kg DM)	7.3	90
Ruminant database parameters		
Metabolizable energy (MJ/kg DM)	7.1	10.9
DM digestibility coefficient	0.48	0.63
Organic matter digestibility coefficient	0.55	0.70

Source: Van Vuuren and Meijs, 1987

material. The different processes can lead to different compositions; for example, related crude protein content varies from 122 g/kg DM for barn-cured hay to 99 g/kg DM for the sun-cured variety (MAFF, 1986). In this case also, information on the type of process applied to the forage is essential in order to maintain the two feeds as distinct entities and, thus, avoid erroneous mean values.

Extrinsic variability of data

Differences in analytical procedures may exist among different laboratories and also within the same laboratory. If not identified, differences can compromise the information retrieved from the database as it will compare values provided on different bases. In this respect, an important limitation of databases is related to the type and quality of the data they make available to their users.

In general, the diversity of analysis methods adopted by different laboratories is responsible for most of the extrinsic variability found in databases (Barber, 1983). It is safe to compare or mean parameters only if they have been grouped according to the method of analysis used to obtain them. An example in which incorrect mean values may result is in the estimation of lignin using potassium permanganate, according to the method of Goering and Van Soest (1970), and using acetyl bromide, according to the method of Morrison (1972). These two procedures for measuring lignin not only give different results but are also expressed in different units and it is therefore incorrect, if not impossible, to merge and compare the two values. Recently, Beever and Mould (2000) highlighted large differences in the crude protein and starch contents of maize silages submitted to nine different laboratories. The values observed for one maize silage ranged from 57 to 119 g/kg DM, for crude protein, and from 165 to 272 g/kg DM, for starch. Investigations uncovered the fact that different methods were being used and that some were predicted by near infrared reflectance spectrophotometry (NIRS).

Variability also often arises in data on the nutritional value of feedstuffs that have been obtained through *in vivo*/*in situ* experiments using animals. The animal species may be a source of this inherent data variability. In particular, it is common practice to classify the nutritional value of feedstuffs on the basis of data obtained through measurement among both sheep and cattle under the general definition of "ruminants". Differences in the passage rate of digesta and, therefore, of feed utilization between these two species are often substantial. Other generalizations commonly made arise from the lack of differentiation among the different physiological states of animals of the same species (dry and lactating cows, for example, or growing and finishing pigs). Such physiological differences may have a great impact on the data obtained, and when not recorded will compromise the quality of those data. Moreover, the use of non-standardized feeding and faecal collection protocols can affect the results obtained substantially.

The *in situ* method used extensively to measure the rumen degradability of feeds is subject to many sources of variability. Huntington and Givens (1995) reviewed many of these, and the *in situ* method has been the subject of a number of ring tests. For example, Madsen and Hvelplund (1994) reported on the protein degradability of five concentrate feeds evaluated *in situ* at 23 different

European institutes. Although the mean values obtained were similar to published values, the reproducibility among centres was poor. In particular, the ring test underlined the need to standardize the material used to make the bags and the washing procedures followed.

A last source of variability, in terms of mean data, is related to the number of samples taken of a feedstuff which can vary widely, resulting in consolidated data that are more or less representative.

In conclusion, in order to produce a high-quality database on feedstuff composition and nutritional value, the intrinsic variability generally needs to be recorded and differences among materials need to be clearly identified. This is vital if effectively different feeds are not to be merged and averaged but maintained distinct. On the other hand, extrinsic variability should be kept to a minimum because, when a few standard procedures are each applied to a large number of samples, the resulting data will be more representative and useful than when a large number of different methods have each been applied to only a few samples.

4. EFFECTS OF DATA VARIABILITY

The use of information obtained from feedstuff databases characterized by low-quality data (with high extrinsic variability and low intrinsic variability) may have various effects depending on the end user of the database (e.g. animal nutritionist, compound feed manufacturer, farm manager, policy-maker). In this respect, in general, the quality of the data is more often influenced by the extrinsic than the intrinsic variability.

For the animal nutritionist

Analysis of feedstuffs composition is costly and time-consuming. Nutritional values acquired through studies with animals are particularly labour-intensive. The nutritionist can therefore not normally afford to analyse specifically each of the feedstuffs involved and often needs to base assessments of nutritive value on more readily available and less expensive information.

The animal nutritionist is in charge of designing the animals' diet. The first consequence of unreliable information on feedstuffs, resulting from high extrinsic variability, may be the formulation of a nutritionally unbalanced diet. This may have different effects, depending on nutritional, environmental and economical factors; these effects are highly interrelated since the optimization of feed utilization leads to the optimization of animal output and financial returns.

Nutritional effects. Today, nutritional information on feedstuffs is used to predict animal performance (often with the use of complex models) through estimation of the energy supply to the animal. The incorporation of low-quality data in these models will produce unreliable estimates of animal performance. Furthermore, it is important to maintain the data resulting from in vivo measurement of nutritive value separate from those values predicted from chemical composition.

Environmental effects. The most common environmental effect derived from unbalanced diets is caused by nitrogen loss. Particularly for ruminants, the capture of nitrogen depends on the rate of production of microbial protein and is related to the availability of carbohydrates and energy in the rumen. It is therefore important to know the exact amount of degradable nitrogen (together with the carbohydrate and energy availability) when characterizing a feedstuff, in order to capture nitrogen for anabolic purposes and avoid the release of ammonia into the environment.

Table 2 gives an example of how the efficiency of nitrogen conversion in the rumen is affected by the level of nitrogen in the diet. In the Table 2 example, the differences in nitrogen intake are due only to differences in the crude protein content of the grass part of the diet. The balance among nitrogen absorption in the intestine, its conversion into milk protein and its loss in faeces and urine is highly dependent on the capacity of the rumen to capture this nutrient. In the example, increasing the nitrogen intake by approximately 25 percent resulted in nitrogen losses in the urine increasing by 80 percent. It is therefore evident that even small differences in the concentration of nitrogen in different feedstuffs may give rise to substantial losses of nitrogen to the environment.

TABLE 2
Nitrogen (N) balance of cows fed at different N levels

N intake	N excretion (g/day)		
(g/day)	Milk	Faeces	Urine
626	107	158	361
494	118	178	198

Source: Van Vuuren and Meijs, 1987.

Economic effects. If, for example, maize gluten is fed at the rate of 5 kg per cow per day to a herd of 100 cows at a metabolizable energy (ME) level of 1.8 MJ/kg DM (as proposed by AFRC, 1993), it would supply the herd with 5 310 MJ ME/day. At a level of 12.8 MJ/kg DM (MAFF, 1986), the daily ME supply to the herd would be 5 760 MJ. Over a 200-day period, the difference in ME supply would be 90 000 MJ, which represents a saving of some 8 tonnes of compound feed, or some US$2 500.

Unbalanced diets may also produce an economic loss in terms of animal health, feed conversion efficiency and, ultimately, the output of animal products. The wrong proportion of a nutrient in the diet is reflected in milk production (in terms of both quantity and quality) and has a direct effect on the economics of production. Such effects can be seen extremely rapidly in fast-growing poultry, where even small changes in the nutrient balance can lead to drastic economic consequences.

For the compound feed manufacturer

Compound feeds are composed of a blend of various raw materials and/or by-products. In designing these products, the feed manufacturer needs to find the balanced combination among different materials that will satisfy the requirements of specific animals in a specific physiological state. In order to achieve this, the manufacturer initially needs to base purchasing decisions on the available information about the chemical composition and nutritional value of feedstuffs. It is therefore very important at this stage that the information available is representative and reliable, not least so that the economic value of the material can be assessed and compared with its market price.

Subsequently, the composition of the different ingredients will be measured and the nutritional value of the product estimated, but failure to get the initial values correct will ultimately be reflected in a higher cost of production and will reduce sales of compound feeds to farmers.

For the farm manager

The farm manager needs to plan crop production in order to meet the animals' requirements. Managers therefore need to obtain information on the supposed chemical and nutritional characteristics of various ingredients in advance and, based on this information, plan the correct combination and quantity of different forages. If the information is inaccurate, the crops produced will provide unbalanced proportions of nutrients (i.e. too much protein, too few available carbohydrates, etc.). The immediate consequences will be an inappropriate diet for the animals, with direct effects on production, and the necessity to compensate the lack of some nutrients by purchasing from the market.

For the policy-maker

The policy-maker needs reliable information in order to be able to direct the policy of the country or region towards competitive and sustainable agriculture. It is evident that, if estimates are based on imprecise information, policy-makers will not be able to make informed or useful decisions.

A further effect of the use of low-quality data by policy-makers is related to the intrinsic variability of feeds, and therefore to real differences in characterizing the materials (genetic differences or differences arising from the processing of raw materials). These differences may be derived from the various ways of describing and naming feedstuffs adopted in different countries. The result is confusion in the trading of feeds, and difficulties in developing and applying export policies.

The lack of harmonization of feed description and naming has been the source of many problems among countries and, in particular, among the Member States of the European Union (EU). Historically, both linguistic and scientific barriers have produced significant differences in feed identification, feed naming and feed description, and false variability often exists because the same name is attributed to different feeds or different names are attributed to identical feeds. The availability of reliable feed data and information that are uniform among different countries will not only contribute to fair trade in feedstuffs but will also facilitate cooperation in animal nutrition research directed towards feed efficiency, product quality and the environment. Not least, it is indispensable for the correct application of the General Agreement on Tariffs and Trade (GATT) and realization of the EU's agricultural policy. Within the EU, a number of steps have been taken to produce a common approach to feed description and naming (ENFIC, 1996).

5. CONCLUSIONS

Feedstuffs vary because of their genetic make-up and as a consequence of the processes applied to them (intrinsic variability of feedstuffs). It is important that information about these factors are available and that genetically different feeds and those processed under varying conditions are identified and recorded separately. Failure to do so will produce improperly averaged, generalized data that provide meaningless information. Another source of data variability results from differences in the methodologies used to obtain the information (extrinsic variability of feedstuffs). Chemical analysis procedures and animal study protocols may vary according to the laboratory or institute involved. This type of variability needs to be minimal if reliable information is to be obtained.

To achieve the necessary reliability, information regarding the chemical and nutritional characteristics of a feedstuff needs to be carefully examined before it is incorporated into a database and, once incorporated, it needs to be carefully managed. Failing to do so will produce false variability among feeds, ultimately resulting in errors in predicting animal performance and environmental effects and impairing the economics of animal products.

REFERENCES

AFRC. 1993. *Energy and protein requirements of ruminants. An advisory manual.* AFRC Technical Committee on Responses to Nutrients. Wallingford, UK, CAB International. p. 136.

Barber, W.P. 1983. Data quality: how it is assessed and improved and what affects it. *In* G.E. Robards & R.G. Packham, eds. *Feed information and animal production*, pp. 57–78. Commonwealth Agricultural Bureaux, Farnham Royal, UK.

Beever, D.E. & Mould, F.L. 2000. Forage evaluation for efficient ruminant livestock production. *In* D.I. Givens, E.R. Owen, R.F.E. Axford & H.M. Omed, eds. *Forage evaluation in ruminant nutrition*, pp. 15–42.

Wallingford, UK. CABI Publishing. **ENFIC.** 1996. EU concerted action animal feed and nutrition. European Network of Feed Information Centres (ENFIC) at Internet address: home.wxs.nl/~enfic/

Fisher, C. 1983. Acquisition and assessment of data on the amino acid composition of feedingstuffs. *In* G.E. Robards and R.G. Packham, eds. *Feed information and animal production*, pp. 285-289. Commonwealth Agricultural Bureaux, Farnham Royal, UK.

Givens, D.I., Cottyn, B.G., Dewey, P.J.S. & Steg, A. 1995. A comparison of the neutral detergent-cellulase method with other laboratory methods for predicting the digestibility *in vivo* of maize silage from three European countries. *Animal Feed Science and Technology*, 54: 55=-64.

Goering, H.K. & Van Soest, P.J. 1970. Forage fibre analysis. *Agriculture Handbook* No. 379. Washington, DC, United States Department of Agriculture (USDA).

Harris, L.E., Asplund, J.M. & Crampton, E.W. 1968. An international feed nomenclature and methods for summarising and using feed data to calculate diets. *Utah State University Agricultural Experiment Station Bulletin* No. 479, November 1968.

Huntington, J.A. & Givens, D.I. 1995. The *in situ* techniques for studying the rumen degradation of feeds: a review of the procedure. *Nutrition Abstracts and Reviews*, (series B) 2(65): 64–93.

Leche, T. 1983. How data are acquired by feed information centres: problems and solutions. *In* G.E. Robards and R.G. Packham, eds. *Feed information and animal production,* pp. 34–36. Commonwealth Agricultural Bureaux, Farnham Royal, UK.

Madsen, J. & Hvelplund, T. 1994. Prediction of *in situ* protein degradability in the rumen. Results of a European ring test. *Acta Agric. Scand.*, Suppl. 25: 103–124.

MAFF. 1986. *Feed composition: UK tables of feed composition and nutritional value*. Marlow Bottom, UK, Chalcombe Publications for the Ministry of Agriculture, Fisheries and Food (MAFF).

Morrison, I.M. 1972. A semi-micro method for the determination of lignin and its use in predicting the digestibility of forage crops. *Journal of the Science of Food and Agriculture,* 23: 455–463.

Thaer, A. 1809. *Grundsatze der rationelle Landwirtshaft*. Vol. 1 Section 275. Berlin, Germany, die Realschulbuchhandlung.

Topps, J.H. 1989. Databases of feed composition and nutritive value. *In* B.A. Stark, J.M. Wilkinson and D.I. Givens, eds. *Ruminant feed evaluation and utilisation*, pp. 41–50. Marlow Bottom, UK, Chalcombe Publications.

Tran, G. & Lapière, O. 1998. *The French feed database: a national network*. Paris, INAPG.

Van Vuuren, A.M. & Meijs, J.A.C. 1987. Effects of herbage composition and supplement feeding on the excretion of nitrogen in dung and urine by grazing dairy cows. *In* H.G. van der Meer, R.J. Unwin, T.A. van Dijk and G.C. Ennik, eds. *Animal manure on grassland and fodder crops. Fertilizer or waste?*

Recent advances in the *in vitro* gas method for evaluation of nutritional quality of feed resources

Harinder P.S. Makkar
Animal Production and Health Section
Joint FAO/IAEA Division
International Atomic Energy Agency
Vienna, Austria

This paper highlights the potential of a novel approach using an *in vitro* rumen fermentation technique for evaluation of nutritional quality of feed resources. This technique enables selection of feed or feed constituents for high efficiency of microbial protein synthesis in the rumen along with high dry matter digestibility, and provides a basis for development of feeding strategies to maximise substrate fixation into microbial cells. This could lead to increase in the supply of protein to intestine and reduce methane production from ruminants. In addition, this technique provides an easy tool to study the effects of various plant active moieties or synthetic compounds for their adverse or beneficial effects on partitioning of nutrients to fermentative gases, short chain fatty acids, and microbial mass.

1. INTRODUCTION

A major constraint to livestock production in developing countries is the scarcity and fluctuating quantity and quality of the year-round feed supply. Providing adequate good-quality feed to livestock to raise and maintain their productivity is and will be a major challenge to agricultural scientists and policy makers all over the world. Increase in population and rapid growth in world economies will lead to increase in demand for animal products; an increase of approximately 30 percent in both meat and milk production is expected in the coming 20 years. At the same time, the demand for food crops will also increase. Future hopes of feeding the millions and safeguarding their food security will depend on the enhanced and efficient utilisation of unconventional resources, which can not be used as food for humans, as feed for livestock. In addition, a large area of land in the world is degraded, barren or marginal and the amount is increasing every year. This also calls not only for identification and introduction of new and lesser-known plants capable of growing in poor soils, which can play a vital role in the control of soil erosion in addition to providing food and feed. In developing countries, livestock are fed mainly on agro-industrial by-products containing a larger proportion of ligno-cellulosic feeds like cereal straws, stovers, sugarcane by-products and similar other feeds. These feeds are poor in protein, energy, minerals and vitamins. Addition of foliage from tree leaves or supplementation with seed meals or even urea can improve the utilization of low quality roughages mainly through the supply of nitrogen to rumen microbes. The use of simple but robust techniques for evaluation of the nutritional quality of these feed resources will contribute to their efficient utilization.

Both growth and milk yield of ruminants are largely limited by forage quality which is mainly reflected in low voluntary intake and digestibility. The importance of these parameters in animal nutrition has long been recognised. The determination of intake and digestibility of feedstuffs *in vivo* is time-consuming, laborious, and expensive, requires large quantities of feed and is unsuitable for large-scale feed evaluation. Therefore, many attempts have been made to predict intake and digestibility using laboratory techniques. Much effort has been directed towards the development of regression equations to predict digestibility from forage chemical composition, but a regression equation that satisfactorily predicts a wide range of forages has not yet been derived.

This paper highlights the potential of a novel approach using an *in vitro* rumen fermentation technique for evaluation of the nutritional quality of conventional and unconventional feed resources.

2. EVALUATION OF FEED RESOURCES

Recent advances in ration balancing include manipulation of feed to increase the quantity and quality of protein and energy delivered to the small intestine. Selection of fibrous feeds based on high efficiency of microbial protein synthesis in the rumen along with high dry matter digestibility, and development of feeding strategies based on high efficiency as well as high microbial protein synthesis in the rumen will lead to higher supply of protein post-ruminally. This concept of feed evaluation has an extra element of the efficiency of microbial protein synthesis in addition to the more conventional one of the dry matter digestibility. The limited supply of protein post-ruminally under most feeding systems in developing countries is an important limiting factor which prevents an increase in animal productivity.

There are a number of methods used to determine net microbial protein synthesis in the rumen based on the use of microbial markers. They require the use of post-ruminally cannulated animals to determine flow of digesta. The cannulation approach is tedious and has several limitations (Chen *et al.*, 1990) to its applicability under most research conditions in developing countries. A simpler technique for determination of microbial protein supply to the intestine is based on the determination of total urinary purine derivatives (IAEA, 1997). This approach is being thoroughly investigated under a Joint FAO/IAEA Coordinated Research Project (IAEA, 1998). Although the method is based on the collection of urine for determination of purine derivatives (allantoin and uric acid for cattle, and allantoin, uric acid, xanthine and hypoxanthine for sheep), the approach is being further simplified using spot urine samples. This technique does not require cannulated animals, but it involves feeding the diets under investigation to animals and therefore is not suitable for screening large numbers of samples or for developing feed supplementation strategies using various feed constituents.

In vitro methods

In vitro methods for laboratory estimations of degraded feeds are important for ruminant nutritionists. An efficient laboratory method should be reproducible and should correlate well with actually measured *in vivo* parameters. *In vitro* methods have the advantage not only of being less expensive and less time-consuming, but they allow maintaining experimental conditions more precisely than do *in vivo* trials. Three major biological digestion techniques are currently available to determine the nutritive value of ruminant feeds: 1) digestion with rumen micro organisms as in Tilley and Terry (1963) or using a gas method (Menke *et al.*, 1979), 2) *in situ* incubation of samples in nylon bags in the rumen (Mehrez and Orskov, 1977), and 3) cell-free fungal cellulase De Boever *et al.*, 1986). These biological methods are more meaningful since microorganisms and enzymes are more sensitive to factors influencing the rate and extent of digestion than are chemical methods (Van Soest, 1994). The nylon bag technique has been used for many years to provide estimates of both the rate and extent of disappearance of feed constituents. This technique provides a useful means to estimate rates of disappearance and potential ruminal degradability of feedstuffs and feed constituents whilst incorporating effects of particulate passage rate from the rumen. The disadvantage of the method is that only a small number of forage samples can be assessed at any one time, and it also requires at least three fistulated animals to account for variations due to animal. It is therefore of limited value in laboratories undertaking routine screening of a large numbers of samples. It is also laborious, and requires large amounts of samples. Substantial error could result in values obtained at early stages of digestion due to a low weight loss; and for poor quality roughages, adherence of microbes at early stages can even lead to higher weights and thus distortion of results if kinetic modelling does not incorporate the lag-phase (McDonald, 1981; Dhanoa, 1988).

The Tilley and Terry (1963) technique is used widely because of its convenience, particularly when large-scale testing of feedstuffs is required. This method is employed in many forage evaluation laboratories and involves two stages in which forages are subjected to 48 h fermentation in a buffer solution containing rumen fluid, followed by 48 h of digestion with pepsin in an acid solution. The method was modified by Goering and Van Soest (1970), in that the residue after 48 h incubation was treated with neutral detergent solution to estimate true dry matter digestibility. Although the method of Tilley and Terry (1963) has been extensively validated with *in vivo* values (Van Soest, 1994), the method appears to have several disadvantages. The method is an end-point measurement (gives only one observation) thus, unless lengthy and labour-intensive time-course studies are made, the technique does not provide information on the kinetics of forage digestion; the residue determination destroys the sample and therefore a large number of replicates are needed. The method is therefore difficult to apply to materials such as tissue culture samples or cell-wall fractions.

Both the Tilley and Terry and nylon bag methods are based on residue determinations and may result in overestimation of dry matter digestibility for tannin-rich feeds, since tannins are solubilised in both these systems but might be indigestible and do not contribute to nutrient supply to animals (Makkar *et al.*, 1993).

In vitro gas method

The gas measuring technique has been widely used for evaluation of nutritive value of feeds. More recently, the increased interest in the efficient utilisation of roughage diets has led to an increase in the use of this technique due to the advantage in studying fermentation kinetics. Gas measurement provides a useful data on digestion kinetics of both soluble and insoluble fractions of feedstuffs. Several gas measuring techniques and *in vitro* gas methods are in use by several groups. Advantages and disadvantages of these methods are discussed by Getachew *et al.* (1998a). The *in vitro* gas method based on syringes (Menke *et al.*, 1979; Blümmel *et al.*, 1997) appears to be the most suitable for use in developing countries. Other *in vitro* methods such as Tilley and Terry and nylon bag methods are based on gravimetric measurements which follow disappearance of the substrate (the components which may or may not necessarily contribute to fermentation), whereas gas measurement focuses on the appearances of fermentation products (soluble but not fermentable products do not contribute to gas production). In the gas method, kinetics of fermentation can be studied on a single sample and therefore a relatively small amount of sample is required or a larger number of samples can be evaluated at a time. The *in vitro* gas method is more efficient than the in sacco method in evaluating the effects of tannins or other anti-nutritive factors. In the in sacco method these factors are diluted in the rumen after getting released from the nylon bag and therefore do not affect rumen fermentation appreciably. In addition, the *in vitro* gas method can better monitor nutrient-antinutrient and antinutrient-antinutrient interactions (Makkar *et al.*, 1995a).

A simple *in vitro* approach is described below which is convenient and fast, and allows a large number of samples to be handled at a time. It is based on the quantification of substrate degraded or microbial protein produced using internal or external markers and of gas or short chain fatty acid (SCFA) production in an *in vitro* rumen fermentation system based on syringes (Menke *et al.*, 1979). This method does not require sophisticated equipment or the use of a large number of animals (but one or preferably two fistulated animals are required) and helps selection of feeds or feed constituents based not only on the dry matter digestibility but also on the efficiency of microbial protein synthesis.

In the method of Menke *et al.* (1979), fermentations are conducted in 100 ml capacity calibrated glass syringes containing feedstuff and a buffered rumen fluid. The gas produced on incubation of 200 mg feed dry matter after 24 h of incubation together with the levels of other chemical constituents are used to predict digestibility of organic matter determined *in vivo* and metabolizable energy.

For roughages, the relationships are:

ME (MJ / Kg DM) = 2.20 + 0.136 Gp + 0.057 CP, R2= 0.94

OMD (percent) = 14.88 + 0.889 Gp + 0.45 CP + 0.0651 XA, R2=0.92

Where ME is the metabolizable energy; DM dry matter, OMD organic matter digestibility; CP, crude protein in percent; XA, ash in percent; and Gp, the net gas production in ml from 200 mg dry sample after 24 h of incubation and after correction for the day-today variation in the activity of rumen liquor using the Hohenheim standard.

Aiple *et al.* (1996) compared three laboratory methods (enzymatic, crude nutrient and gas measuring technique) as predictors of net energy (as estimated by equations based on *in vivo* digestibility) content of feeds and found that for predicting net energy content of individual feeds, the gas method was superior to the other two methods.

The method of Menke *et al.* (1979) was modified by Blümmel and Ørskov (1993) in that feeds were incubated in a thermostatically controlled water bath instead of a rotor in an incubator. Makkar *et al.* (1995b) and Blümmel *et al.* (1997) modified the method further by increasing the amount of sample from 200 to 500 mg and increasing the amount of buffer two-fold as a result the incubation volume increase from 30 ml in the method of Menke *et al.* (1979) to 40 ml in the modified method. In the 30 ml system, the linearity between the amount of substrate incubated and the amount of gas produced is lost when the gas volume exceeds 90 ml because of the exhaustion of buffer of the medium; and in 40 ml system, the linearity is lost when the gas volume exceeds 130 ml (Getachew *et al.*, 1998b). The exhaustion of the buffer decreases pH of the incubation medium; consequently the fermentation is inhibited. If the amount of gas production exceeds 90 ml using the 30 ml system and 130 ml using the 40 ml system, the amount of feed being incubated should be reduced.

The main advantages of the modified method (the 40 ml system and incubation in a water bath) are: i) there is only a minimum drop in temperature of the medium during the period of recording gas readings on incubation of syringes in a water bath. This is useful for studying the kinetics of fermentation where gas volume must be recorded at various time intervals, ii) because of large volume of water in the water bath and also its higher temperature holding capacity, drastic drop in the temperature of the incubation is prevented in case of power breakdown for a short duration, and iii) an increase in amount of sample from 200 to 500 mg reduces the inherent error associated with gravimetric determination needed to determine concomitant *in vitro* apparent and true digestibility (see below).

When a feedstuff is incubated with buffered rumen fluid *in vitro*, the carbohydrates are fermented to produce short chain fatty acids, gases and microbial cells. Gas production is basically the result of fermentation of carbohydrates to acetate, propionate and butyrate. Gas production from protein fermentation is relatively small as compared to carbohydrate fermentation. The contribution of fat to gas production is negligible. When 200 mg of coconut oil, palm kernel oil and/or soybean oil were incubated, only 2.0 to 2.8 ml of gas were produced while a similar amount of casein and cellulose produced about 23.4 ml and 80 ml gas.

The gas produced in the gas technique is the direct gas produced as a result of fermentation and the indirect gas produced from the buffering of SCFA. For roughages, when bicarbonate buffer is used, about 50 percent of the total gas is generated from buffering of the SCFA and the rest is evolved directly from fermentation. At very high molar propionate the amount of CO2 generated from buffering of SCFA is about 60 percent of total gas production. Each mmol of SCFA produced from fermentation releases 0.8-1.0 mmol of CO2 from the buffered rumen fluid solution, depending on the amount of phosphate buffer present. Highly significant correlation has been observed between SCFA and gas production (see below).

Gas is produced mainly when substrate is fermented to acetate and butyrate. Substrate fermentation to propionate yields gas only from buffering of the acid and, therefore, relatively lower gas production is associated with propionate production. The gas that is released with the generation of propionate is only the indirect gas produced from buffering. The molar proportion of different SCFA produced is dependent on the type of substrate. Therefore, the molar ratio of acetate of propionate has been used to evaluate substrate related differences. Rapidly fermentable carbohydrates yield relatively higher propionate as compared to acetate, and the reverse takes place when slowly fermentable carbohydrates are incubated. Many workers reported more propionate and thus lower acetate to propionate ratio in the ruminal fluid of cows fed a high grain diet. If fermentation of feeds leads to a higher proportion of acetate, there will be a concomitant increase in gas production compared with a feed with a higher proportion of propionate. In other words, a shift in the proportion of SCFA will be reflected by changes in gas production.

The gas produced on incubation of cereal straws (Blümmel and Ørskov, 1993), a wide range of feeds including many dairy compound feeds and their individual feed components whose protein and fat contents vary greatly (Blümmel *et al.*, 1999a), and tannin containing browses (Getachew *et al.*, 2000a) in absence or presence of polyethylene glycol (a tannin complexing agent) in the buffered rumen fluid was closely related to the production of short chain fatty acids as per Wolin (1960) stoichiometry, which is as follows:

$$\text{Fermentative } CO_2 = A/2+P/4+1.5B$$

where A, P and B are moles of acetate, propionate, and butyrate respectively.

$$\text{Fermentative } CH_4 = (A+2B)-CO_2$$

where A and B are moles of acetate and butyrate respectively and CO_2 is moles of CO_2 calculated from previous equation.

Assumption: one mole of SCFA releases one mole of CO_2 from bicarbonate-based buffer described as buffering CO_2 and therefore, mmol of buffering CO_2 is equal to mmol of total SCFA generated during incubation.

$$\text{Gas volume} = \text{mmol of gas} \times \text{gas constant (R)} \times T$$

where:
R = the ratio between molar volume of gas to temperature (Kelvin zero; K) i.e. (22.41l/273 = 0.082),

T = incubation temperature (Kelvin); 273 + 39o C = 312 K
Total volume of gas (ml) calculated from SCFA production = (BG + FG) x CF
BG = gas volume (ml) from buffering of SCFA,
FG = fermentative gas (ml) ($CO_2 + CH_4$),
CF = correction factor for altitude and pressure which is 0.953 for Hohenheim at altitude 400m above sea level (Blümmel *et al.*, 1999). (The volume of 1 mmol of gas at 39o C in Hohenheim would be; 1 x 0.082 x 312 x 0.953=24.4 ml).

The origin and stoichiometry of gas production have been described in details in Blümmel *et al.* (1997) and Getachew *et al.* (1998a).

The *in vitro* gas production measured after 24 h incubation of tannin containing browses in the presence or absence of polyethylene glycol (PEG) was strongly correlated with gas volume stoichiometrically calculated from SCFA. The relationship between SCFA production (mmol) and gas volume (ml) after 24 h of incubation of browse species with a wide range of crude protein (5.4-27 percent) and phenolic compounds (1.8-25.3 percent and 0.2-21-4 percent total phenols and total tannins as tannic acid equivalent respectively) was (Getachew *et al.*, 2000a):

In the absence of PEG: SCFA = 0.0239.Gas -0.0601; R^2 = 0.953; N = 39; P<0.001 (I)

In the presence of PEG: SCFA = 0.0207.Gas + 0.0207; R^2 = 0.925; N = 37; P<0.001 (II)

These relationships are similar to that obtained for wheat straw (Blümmel *et al.*, 1993).

The SCFA production could be predicted from gas values using the above relationship. The level of SCFA is an indicator of energy availability to the animal. Since SCFA measurement is important for relating feed composition to production parameters and to net energy values of diets, prediction of SCFA from *in vitro* gas measurement will be increasingly important in developing countries where laboratories are seldom equipped with modern equipments to measure SCFA.

The stoichiometric balance also allows calculation of the CH4 and CO2 expected from the rumen fermentation if the molar proportions and amount of SCFA are known.

Kinetics of fermentation feedstuffs can be determined from fermentative gas and the gas released from buffering of SCFA. Kinetics of gas production is dependent on the relative proportion of soluble, insoluble but degradable, and undegradable particles of the feed. Mathematical descriptions of gas production profiles allows analysis of data, evaluation of substrate- and media-related differences, and fermentability of soluble and slowly fermentable components of feeds. Various models have been used to describe gas production profiles. France *et al.* (1993; 2000) combine modelling of the gas profiles with estimates of substrate loss and ruminal rate of passage and derive estimates of ruminal extent of degradation thus linking gas production to events in the rumen proper.

In vitro gas production with concomitant microbial mass measurement

Determination of microbial mass. *In vitro* gas tests are attractive for ruminant nutritionists since it is very easy to measure the volume of gas production with time, but the measurement of gas only implies the measurement of nutritionally wasteful and environmentally hazardous products. In most studies the rate and extent of gas production has been wrongly considered to be equivalent to the rate and extent of substrate (feed) degradation. Current nutritional concepts aim at high microbial efficiency, which cannot be achieved by measurement of gas only. *In vitro* gas measurements reflect only SCFA production. The relationship between SCFA and microbial mass production is not constant and the explanation for this resides in the variation of biomass production per unit Adenosin-Three Phosphate (ATP) generated. This can impose an inverse relationship between gas volume (or SCFA production) and microbial mass production particularly when both are expressed per unit of substrate truly degraded. This implies that selecting roughages by measuring only gas using *in vitro* gas methods might result in a selection against the maximum microbial mass yield. Blümmel *et al.* (1997) have demonstrated how a combination of *in vitro* gas production measurements with a concomitant quantification of the truly degraded substrate provides important information about partitioning of fermentation products, and the *in vitro* microbial mass production can be calculated as:

Microbial mass (mg) = mg substrate truly degraded - (ml gas volume x stoichiometrical factor)

For roughages, the stoichiometrical factor was 2.20.

Partitioning factor. The parameters in the above equation also allow the calculation of a partitioning factor (PF). The PF is defined as the ratio of substrate truly degraded *in vitro* (mg) to the volume of gas (ml) produced by it (equivalent to the reciprocal of parameter Y in France *et al.*, 1993). The above equation becomes

Microbial mass (units) = gas volume (PF - stoichiometrical factor).

A feed with higher PF means that proportionally more of the degraded matter is incorporated into microbial mass, i.e., the efficiency of microbial protein synthesis is higher. Roughages with higher PF have been shown to have higher dry matter intake. Different *in vitro* PF values are also reflected by *in vivo* microbial protein synthesis as estimated by purine derivatives (the higher the PF, the higher the excretion of urinary purine derivatives; (Blümmel *et al.*, 1999b) and in methane production by ruminants (the higher the PF, the lower the methane output; (Blümmel *et al.*, 1999b). These results show that the PF calculated *in vitro* provides meaningful information for predicting the dry matter intake, the microbial mass production in the rumen, and the methane emission of the whole ruminant animal.

The procedures for the determination of truly degraded substrate and the calculation of the stoichiometrical factor; stoichiometrical relationship between SCFAs and gas volume; and relationship between SCFA production, ATP production and microbial mass yield can be obtained from Blümmel *et al.* (1997) and Getachew *et al.* (1998a). It may be noted that these procedures and relationships are valid for substrates consisting predominantly of structural carbohydrates, and the findings might not extend to substrates such as those high in soluble carbohydrate, protein or fat. Rymer and Givens (1999) have shown that, as observed by Blümmel *et al.* (1997), good quality feeds (grass silage, wheat, maize, molasses sugarbeet feed and fishmeal) which produce large amounts of gas and SCFA yield small amounts of microbial mass per unit of feed truly degraded.

It seems therefore justified to suggest that feeds or feed ingredients should be selected that have a high *in vitro* true degradability but low gas production per unit of truly degraded substrate. Dijkstra *et al.* (2000) have described modelling of microbial protein synthesis *in vivo* from the *in vitro* gas parameters.

Incubation time and PF. Another study by Blümmel *et al.* (1998a), in addition to once again describing the importance of measuring microbial mass, has highlighted the importance of the fermentation time at which the microbial mass should be measured. In this study, substrate degradation and kinetics of *in vitro* partitioning of three hays, with similar *in vivo* digestibilities, into SCFA, microbial mass yield, and ammonia, carbon dioxide and methane production was examined at 8, 12, 18 and 24 h of incubation in the gas method under both low and adequate nitrogen levels. Microbial synthesis was quantified gravimetrically (Blümmel *et al.*, 1997), by nitrogen balance (Getachew *et al.*, 2000b) and by purine analysis (Makkar and Becker, 1999). SCFA and gas production were positively correlated ($P < 0.0001$) and cumulative at all times of incubation under both low and adequate nitrogen levels. On the other hand, microbial mass, microbial nitrogen and microbial purine yields declined after 12 h of incubation while ammonia production increased. Per unit of substrate degraded, gas and SCFA production were always inversely ($P < 0.05$) related to microbial mass yield regardless of incubation time and medium (low or adequate nitrogen). At later incubation times, continuously more SCFA or gas and less microbial mass were produced reflecting microbial lysis and probably increasing microbial energy spilling. All three hays differed ($P < 0.05$) consistently in how the degraded substrate was partitioned into SCFA and gas and into microbial mass in both the low and adequate nitrogen medium. Purine analysis indicated substantial differences in microbial composition across treatments, which might be one explanation for these different microbial efficiencies (Blümmel *et al.*, 1998a).

The efficiency of microbial growth was higher for 16 h incubation than 24 h for tannin-rich feeds when these were incubated in presence or absence of PEG, a tannin-inactivating agent. Additional nitrogen in the medium also affected the efficiency of microbial protein synthesis from tannin-tannins feeds both at 16 h and 24 h (Getachew *et al.*, 2000c). For proper characterisation of feed and feed ingredients, approaches need to be developed for measuring the PF at the incubation time at which the lysis of microbes is minimal. Some possible simple approaches worth investigating to identify this incubation time are: i) the time at which half of the maximum gas is produced, and ii) the inflection point at which the rate of gas production is maximum (the rate increases up to a certain incubation time and thereafter decreases as the incubation progresses), both these parameters can be mathematically calculated from the gas profiles. The effect of the nitrogen level in the medium on the PF and the significance *in vivo* of these PFs also need to be investigated.

Digestion kinetics of neutral detergent-soluble fraction. The gas measurement method has also been used to study digestion kinetics of the neutral detergent soluble fraction of forages, starch-rich feeds and other highly digestible carbohydrate components, which was obtained by subtracting the average gas production curve for the digestible neutral detergent fibre (NDF) from that of the unfractionated whole feed (Chen *et al.*, 1999). The subtraction procedure might give some useful information relevant to low-NDF fibre feeds, e.g., corn grain Chen *et al.*, 1999), but it is not suitable for forages rich in NDF (Blümmel *et al.*, 1998b). Blümmel *et al.* (1998b) examined the rate and extent of fermentation of whole roughage and extracted NDF, dry matter degradability of extracted NDF and the PF for whole roughage and the extracted NDF of 54 roughages. The 24-h degradabilities of extracted NDF were higher than NDF degradabilities in whole roughages, and the PF values were lower for extracted NDF than for whole roughages (2.5 vs 3.1; i.e. the efficiency of microbial protein synthesis with extracted NDF was lower). Both the higher degradability and lower PF contributed to higher gas volumes obtained from extracted NDF compared with whole roughage. Supplementation of amino acids and sugars, which essentially constitute the solubles, may increase the efficiency of microbial synthesis from cell walls during fermentation (a situation similar to that in unfractionated forages) and the removal of solubles may result in lower microbial efficiencies. A considerable effect of cell solubles on partitioning of nutrients from the NDF raises doubts as to the significance *in vivo* of the kinetic parameters calculated using the subtraction procedure (Blümmel *et al.*, 1998b).

Voluntary intake prediction. The main constraint to the utilisation of roughages by ruminants is voluntary feed intake so prediction of feed intake, particularly of fibrous roughage, is one of the important aspects of ruminant nutrition. *In vitro* gas production has been used to predict dry matter intake. Various workers have reported significant correlation between *in vitro* gas production and dry matter intake. Forage cell walls have considerable influence on voluntary feed intake through rumen fill mechanism (Van Soest, 1994). Gas production from extracted neutral detergent fibre was shown to be better correlated to voluntary feed intake than the values obtained from the incubation of whole roughage. The use of various models for intake prediction was investigated and it currently appears that combination of gas volume measurements (4-8 h) with concomitant determination of the amount of substrate degraded (> 24 h) is superior to the models based on kinetics of gas production only. The *in vitro* gas production from NDF explained more (82 percent vs. 75 percent) of the variation in dry matter intake than gas production from whole roughage (Getachew *et al.*, 1998a).

Interaction between dietary constituents. Gas measurement was also employed for evaluation of the interaction between basal and supplementary diets by incubating basal diet and supplementary diet separately as well as in combination and monitoring gas production at different hours of incubation using the pressure transducer system (Sampath *et al.*, 1995). This will indicate the availability of readily fermentable material as a ready energy source, which will stimulate the activity of the rumen micro-organisms which in turn would accelerate the digestion of roughages. These workers, by incubating the basal diet and the supplement, observed a positive interaction in gas production in the early hours of incubation, which according to the authors can be an approach to study the synergetic effects of supplementation. However, it must be pointed out that measurement of gas only, could lead to misleading results (see above) It is suggested to determine microbial mass production in addition to the gas measurement for such studies.

Evaluation of tannin-containing by-products and forage

The need to determine microbial mass using internal or external markers. Unfortunately, the method based on the gas method and the detergent system of fibre analysis to calculate the microbial mass produced for fibrous feeds (the method outlined above) did not work for tannin-rich feeds. The PF for tannin-rich feeds (calculated as mg truly degraded substrate needed to produce one-ml gas) ranged from 3.1 to 16.1 (Getachew *et al.*, 2000c), which is well above the theoretical range of PF (2.75 to 4.41) (Blümmel *et al.*, 1997). The high PF could be due to: a) solubilization of tannins from the feed. These tannins would make no contribution to gas or energy in the system but would contribute to dry matter loss, b) the cell solubles contributing to dry matter loss but not to gas production because the gas production is inhibited by tannins or c) a combination of a) and b). In addition, the appearance of tannin-protein complexes as artefacts in the true residue also makes the gravimetric approach

(Blümmel *et al.*, 1997) of quantification of microbial mass redundant (Makkar *et al.*, 1997a). The presence of tannin-protein complexes in faecal samples (the origin of proteins could be microbes, feed or endogenous secretion from the gastro-intestinal tract) from animals fed tannin-rich forage and their non-removal by the detergent system of fibre analysis (Van Soest *et al.*, 1991) leads to misleading values of fibre and also causes problems in the *in vivo* evaluation of tannin-rich feeds (Degen *et al.*, 1995; Makkar *et al.*, 1995c). Therefore, caution is required in interpreting results obtained from *in vivo* or *in vitro* experiments on the evaluation of tannin-containing feeds using the detergent system of fibre analysis.

For the *in vitro* evaluation of tannin-rich feeds, the microbial mass should be quantified using diaminopimelic acid (DAPA) or purines as markers, or by 15N incorporation into the microbes (Makkar *et al.*, 1998a), and the PF for tannin-rich feeds can be expressed as the microbial mass determined by these markers per ml of gas produced (or per mmol SCFA produced). The system developed for evaluation of tannin-containing feeds depends on incubation of the feed in the presence and absence of polyethylene glycol (PEG MW 4000 or 6000 preferably of 6000; (Makkar *et al.*, 1995b)) and measurement of gas (or SCFA) and microbial mass using the above mentioned markers. PEG has a high affinity for tannins. Addition of PEG results in the formation of PEG-tannin complexes which inactivates tannins. The changes in gas (or SCFA) and microbial mass as a result of PEG addition represents 'in totality' the tannin effects (biological) as a function of the rumen fermentation parameters. This bioassay based on an *in vitro* rumen fermentation system coupled with the use of a tannin-complexing agent, PEG could be complementary to other tannin assays (Makkar *et al.*, 1995c; Makkar, 1989; Makkar, 2000) in evaluating the nutritional quality of tanniniferous feeds.

Significance of bound tannins and the efficiency of microbial protein synthesis. The above approach of incubating tannin-containing feeds in the presence and absence of PEG also enables studies to be made on the nutritional significance of both extractable and unextractable (bound) tannins. Addition of PEG during the incubation of tannin-rich NDF led to an increase in gas production, suggesting that tannins released as a result of NDF degradation by rumen microbes are biologically active and have the potential to influence rumen fermentation (Makkar *et al.*, 1997b). Similar results were obtained on incubation of tannin-rich browses made free of extractable tannins by repeated use of 70 percent aqueous acetone.

Another application of this method is to study the effect of tannins on the partitioning of nutrients between microbial protein and SCFAs or gases or to study the efficiency of rumen microbial protein synthesis. Using DAPA, purines and 15N approaches for measuring microbial mass it has been shown that in the presence of PEG, the degradabilities of substrate and microbial mass production were higher, the efficiency of microbial protein synthesis was lower (Makkar *et al.*, 1995a, 1998a; Getachew *et al.*, 2000b). Similar results have also been obtained by other approaches based on the gas method in which the rate of ammonia uptake is taken as the efficiency of microbial protein synthesis (Makkar *et al.*, 1998a), and using nitrogen balance approach (Getachew *et al.*, 2000b). Conversely, efficiency of microbial protein synthesis is expected to be higher in the presence of tannins. The net microbial mass production would depend on the balance between decreased degradable dry matter and higher microbial mass production per unit of dry matter digested in presence of tannins.

This method has also been used to develop strategies to enhance the microbial mass production through enhancing efficiency of microbial protein synthesis for tannin rich feeds. Incorporation of PEG in small amounts and scattered over the incubation period increased the efficiency of microbial mass production (Getachew *et al.*, 2000d) which translated into the inclusion of PEG in fed blocks. The licking of these blocks by animals resulted in slow release of PEG in the rumen which produced higher animal productivity compared to when the same amount of PEG was given to animals in one portion (Ben Salem *et al.*, 1999).

Protein degradability. Although the 24-h *in vitro* degradable nitrogen values obtained from tannin-rich browse and herbaceous legumes were lower than those reported for low quality roughages (Getachew *et al.*, 1998b; 2000b), the relatively high crude protein content of these browses and herbaceous legumes could play a significant role in supplying rumen degradable nitrogen. In the presence of a tannin-inactivating agent, PEG, their *in vitro* degradable nitrogen values were raised. The difference between these values observed in the presence and absence of PEG indicates the amount of protein protected by tannins from degradation in the rumen (Getachew *et al.*, 2000b). Whether the protein protected by tannins from microbial degradation is fully available to animal's post-ruminally requires further research. Raab *et al.* (1983) reported a close relationship between *in vivo* and *in vitro* values when incubation was terminated after 17 h. When normal proteins feed was tested about 80 percent of the 24 h value was degraded in the first 8 h incubation whereas in protected protein feed only 60 percent of the 24 h value was degraded in this time (Raab *et al.*, 1983). The appropriate incubation time for *in vitro* degradability studies in the presence and absence of PEG for tannin-rich feeds may depend on the nature of protein and tannins, which should be identified. Also, this method for quantifying feed proteins that have been protected by tannins from degradation in the rumen needs to be validated *in vivo*.

Measurement of feed proteins during fermentation using polyacrylaminde gel electrophoresis coupled with the use of an image analyser could be another attractive approach for measuring the rumen degradability of nitrogen, and for studying the influence of various natural plant products (e.g., tannins, saponins, alkaloids; (Makkar *et al.*, 1997a, 1998a,b).

3. SOME PRACTICAL CONSIDERATIONS FOR FEED EVALUATION USING THE GAS METHOD

After 16 h (for concentrate feeds) and 24 h (for roughage samples) of incubation (The termination of the incubation should be at the time at which the efficiency of microbial production is maximum, which is difficult to ascertain under normal feed evaluation experiments for a large number of feeds. An approach that could be applied is running of

200 mg sample for the kinetic experiment and calculating the time (t/2) at which the gas production is half of the potential gas production. The time 't/2' is near to the time at which the efficiency of microbial production is the maximum. This means that one has to run a kinetic experiment before determination of truly degraded organic matter and microbial protein using purines or 15N approaches. The time 16 h for concentrate feeds and 24 h for roughages might be a good compromise for routine feed evaluation), measure:

a) True organic matter degradability and gas production (weigh air dried 375 + 5 mg sample in 30 ml medium containing double the level of bicarbonate ions than the conventional medium of Menke *et al.* (1979) used for estimation of metabolizable energy and degradability of organic matter (DOM)). Record net gas production (ml gas production in syringe containing feed minus ml gas production in blank). After termination of the incubation, transfer the contents of the syringe quantitatively in a beaker and digest it with the neutral digestion solution for 1 h (the purpose of this treatment is to solubilize the microbes and obtain only the under graded feed). Filter the contents through crucible no. 2, wash the residue on the crucible with hot water till the residue is free of the detergent. Dry the crucibles at 130 °C for 2 h or at 100 °C for 10 h (overnight). Record weight of the crucibles after transferring them to a desiccator. This weight minus weight of empty crucible gives the weight of under graded feed in that particular syringe. [Note: this method of measurement of under graded residue does not work satisfactorily with tannin-containing feeds/samples (presence of tannin-protein complexes as artefacts in the residue) and starch-rich feeds (some starch might not be degraded by microbes up to 16 h of incubation but it would get solubilized in the neutral detergent solution leading to underestimation of truly under graded residue). It may be noted that this method of measurement of under graded residue should not be applied at the initial hours of incubation (before 16 h of incubation) since during this period a portion of the feed, which is under graded by microbes could be solubilised in the neutral detergent solution.]. Let the weight of this residue (under graded feed) be 'a'. Now transfer the crucibles containing this residue to Muffle furnace and ash the sample. The organic matter will disappear leaving the ash. After transferring the crucibles to a desiccator, weigh them and subtract from this weight the weight of the empty crucible to obtain the weight of ash (called as 'b'), i.e. subtract 'b' from 'a' to obtain truly under graded organic matter (a-b). Calculate the organic matter degradability in percentage unit. Please note that for calculation of organic matter degradability, one needs to have dry matter and ash contents of the feed sample. This will enable to calculate organic matter in 375 + 5 mg of the air-dried matter weighed in the syringe for the incubation.

Organic matter weighed into the syringe

= (375 x DM in percentage/100)) minus (375 x (DM in percentage/ 100) x (Ash in percentage/100)).

Or (375 x DM in percentage/100) (1 – Ash in percentage/100)

Let this value be 'c'.

Percent organic matter degradability is (a-b)100/c.

b) For most conventional feed resources (not the tannin-containing feeds and starch-rich feed ingredients), microbial mass production can be estimated at the time of termination of the incubation. Please note that this approach should not be applied for the initial hours of incubation (earlier than 16 h).

mg microbial mass production = ((a-b) – 2.2 x net gas in ml)); (a-b) is in mg and 2.2 is the stoichiometric factor.

Efficiency of microbial mass production = ((a-b) – 2.2 x net gas in ml)/(a-b)

c) Calculation of PF. PF is a measure of efficiency of microbial mass production or efficiency of microbial protein production. Higher the factor, higher is the efficiency.

PF = (mg truly degraded organic matter)/ml gas
Or
PF = c – (a-b) in mg truly degraded organic matter divided by ml gas produced in that particular syringe for which 'c – (a-b)' has been calculated. This value is the PF.
The theoretical range for PF is 2.74 to 4.41. Any value above or below this range should be critically evaluated. For tannin-rich samples, the PF is normally above 4.41. The higher PF (for example 7.2) would mean that 7.2 mg of the truly degraded organic matter produce one ml of the gas. The high values for tannin-rich feeds appear to be due to the solubilization of tannins from the feeds during fermentation, contributing to the dry matter loss but without contributing to the gas, and inhibition of gas production from cell solubles by tannins through inhibition of rumen fermentation (cell solubles will also contribute to dry matter loss). The presence of tannin-protein complexes in the truly under graded residue and hence underestimation of the truly degraded organic matter also contribute to the errors in the PF, but the presence of these complexes in the truly unde graded will lower the PF. The PF, which is obtained for the tannin-containing sample is the net effect of solubilization of the feed components during fermentation but not contributing to the gas production and the presence of tannin-protein complexes in the truly undegraded residue.

d) For tannin-containing feeds, the syringe contents are not digested after incubation with the neutral detergent solution (see above) to determine truly undegraded organic matter. Instead, the syringe contents are taken to the determination of purines and/or 15N incorporation studies.
Efficiency of microbial mass production can be expressed as: mg net purines in syringe divided by ml of net gas production (both these values need to be corrected for the blanks).
Or as:

mg (or microgram) of 15N in apparently under graded residue/ml of net gas production (both these values need to be corrected for the blanks).

If need be, purine amount can be converted to microbial-N by taking rumen fluid sample, centrifuging it at approx. 17,000 g to obtain microbial pellet, washing it once with distilled water and recentrifuging, and lyophilising it. In a weighed lyophilised pellet, purines can be determined by high performance chromatography (HPLC) or using a spectrophotometeric method after precipitation of purine with silver nitrate. Using this purine to microbial mass ratio, the purine can be converted into microbial mass. If there is a need to convert purine to microbial-N, a portion of the same lyophilised pellet can be subjected to the determination of N using micro-Kjeldahl method or by converting N in the pellet to ammonia and determination of ammonia using sodium nitroprusside and hypochlorite reaction (Makkar and Becker, 1999).

Note: Purines are only in the microbes and not in the supernatant, but 15N is present both in the apparently undegraded residue as well as in the supernatant after the incubation. Therefore, there is a need to take apparently undegraded residue for 15N incorporation studies. The apparently undegraded residue is prepared by centrifuging the syringe content using the high speed centrifuge (17,000 g for 20 min at 4o C), discarding supernatant, washing the residue with distilled water followed by again the high speed centrifugation and discarding the supernatant. The residue should be lyophilized or can be dried in a vacuum oven at approx. 50o C. The weight of this residue should be determined (weight of centrifuge tube plus the residue) minus weight of empty centrifuge tube. This weight needs to be taken into account in the calculations at a later stage. The residue should be quantitatively collected and ground to fine powder preferably using ball mill. In addition to determination of 15N, purines could also be determined in this residue. This section addresses the use of this approach for tannin-containing feeds but this could be used for any feed samples.

e) The gas production can be converted to mmol SCFA production using the equations (I) and (II).

f) Estimation of *in vivo* metabolizable energy and digestibility of organic matter (DOM) using the net gas value from 375 mg sample and converting the gas value from 375 mg to 200 mg dry sample or directly incubating 200 mg sample for 24 h.

g) Determination of tannin activity by incubating 375 mg of sample in presence and absence of PEG 6000 in 30 ml of the medium containing double the level of bicarbonate ions (see section 'a' above); termination of the incubation at 16 and 24 h; and expressing the results at both 16 and 24 h of the incubation, as:
- Difference in the gas production as percent increase on addition of the PEG compared to the syringe without the PEG,
- Difference in purine as percent increase or decrease on addition of the PEG compared to the syringe without the PEG,
- ME and DOM values with and without PEG (after converting gas value for 200 mg sample), and percent change in ME and DOM values on addition of the PEG (taking the value without PEG as 100 percent), and
- Change in the efficiency of microbial protein production when expressed as mg purine (or mg 15N incorporation)/ml gas, on addition of the PEG.

h) Determination of the kinetics of gas production (France *et al.*, 2000) by incubating 200 mg of the sample for 96 h for roughages and 72 h for concentrate samples.

i) Microbial N (MN) could also be measured after incubation by

following two nitrogen balance approaches (Getachew *et al.*, 2000b). The first approach is:
MN = TN – (NDF-N + Ammonia-N),
where TN is total N i.e., feed N + N in buffered rumen fluid in the syringe before incubation (at 0 time), NDF-N is the N bound to NDF fraction following incubation and Ammonia-N is the ammonia-N in the supernatant following the incubation. In a closed system, the total N present at the start of the incubation can be in microbial mass, NDF-N, ammonia-N and amino acids during any time of the incubation. Negligible amounts of amino acids and peptides are present in the supernatant during fermentation and therefore these can be ignored in calculation of microbial N.
For determination of NDF fraction for N bound to NDF after incubation, the syringe contents are transferred into a 600 ml beaker and the syringes are washed twice with a total of 50 ml neutral detergent solution (NDS) and emptied into the beaker. The contents are refluxed for 1 h, and then filtered through pre-tarred filter crucibles (no. 2). The crucibles are dried overnight at 100 °C and weighed. True degradability can be calculated as the weight of substrate incubated minus the weight of the residue after NDS treatment. The residue after NDS treatment (neutral detergent residue, NDF) is subjected to micro-Kjeldahl digestion for determination of NDF-N.

The second approach is:
MN = APUR-N – NDF-N,
where APUR-N is N bound to apparent under graded residue after incubation. The preparation of apparent under graded residue is given above (section 'd'). N bound to this fraction is determined in a manner similar to NDF-N determination.

From the weight of apparent under graded residue, apparent digestibility can also be calculated.

Do's and don'ts for the gas method

- Soon after getting the syringes from the firm, mark the plunger and the corresponding graduated outer portion (barrel) with a diamond pencil (just after opening the case containing the syringe comprising of the plunger and the barrel). Give the same number to both the parts of the syringe.
- The plunger should be properly lubricated using white Vaseline (apply less amount of Vaseline for incubations up to 24 h and more for incubations up to 96 h).
- Collect rumen liquor from both the liquid and the solid phase and handle it properly (use of warm containers, flushing the containers with carbon dioxide, always keeping the rumen liquor under carbon dioxide).
- Reducing solution should be prepared fresh on the same day of conducting the experiment.
- Start flushing the medium with carbon dioxide well before (approx. 10 min) adding the reducing solution. Also flush the medium for at least 10 min after adding the rumen liquor and before starting filling the syringes. Keep flushing the medium with carbon dioxide while filling the syringes (the flow could be reduced at this stage).
- While filling the syringes with the medium, keep an eye on the medium (carbon dioxide gas should be flushing into the medium and the medium should be stirring).
- After dispensing 30 ml of the medium into the syringe, create a light vacuum by pushing back the plunger and then open the clip. This procedure will bring the medium lying in the nozzle back into the syringe. Otherwise there could be a loss of the medium and/or sample.
- After filling of the syringes has been completed (might take 30-40 min), shake the syringes. Shake them again after every 30 min till first 2 h of the incubation, and then after every two hours till the first 10 or 12 h of the incubation. Thereafter, shake the syringes after taking the gas volume readings (24, 30, 36, 48, 60, 72, 96 h, as the case may be). Make sure that all feed particles are

taken into the medium while stirring (swirling shaking action might help).

- Wash the dispenser with distilled water immediately after finishing filling the syringes, otherwise the dispenser could get stuck up and might not then be usable.
- Check temperature and level of water in the water bath at least twice a day.
- In the evening before going home, if the plunger is above 80 ml level, push it back, record the readings (before and after) pushing back the plunger.
- When you push back the syringe in the evening, give a shake after approximately 30 min in order to prevent taking up the sample along with the bottom portion of the plunger and out from the incubation medium.
- Use the carbon dioxide gas cylinder with caution. Ask someone if you do not know its operation. Misuse could cause an accident.
- While taking the gas volume readings, use the brown ring marked on the plunger and not the bottom end of the plunger. Keep the syringe in inverted position and in parallel with eye while recording the gas reading. Immediately transfer the syringe into the water bath after taking the reading.
- For cleaning the syringes, the syringe should be emptied (preferably pulling back the plunger and removing contents from the back and not from the nozzle). Clips should be removed. The plunger and the outer graduated part of the syringe (barrel) should be separated. Excess Vaseline on the plunger should be cleared with a tissue paper or a piece of soft cloth, and then transfer both the parts in hot water containing detergent (soap) solution. Rub the plunger with hand and inside of the barrel with a soft brush to clean these. Wash thoroughly both the portions with hot water and finally rinse them with distilled water. Dry them well before weighing sample into the syringe.

- Fix the clip in such a manner (by keeping the portion, where pressure is applied to open or close it, facing the syringe) that it does not open by striking on the edges of the lid of the water bath while taking out the syringe for taking reading.
- Mark the crucibles well (preferably with a diamond pencil). Keep them in increasing or decreasing order; this might help you in identifying the crucibles, which have not been marked well, especially after these have been placed in the Muffle furnace.
- For tannin bioassay use only PEG-4000 or PEG-6000 (preferably the latter).
- For a slide show on the gas method, refer to: http://www.iaea.org/programmes/nafa/d3/mtc/invitro-slideshowapr01.pdf

4. CONCLUSIONS AND FUTURE RESEARCH

The *in vitro* rumen fermentation method in which gas production and microbial mass production are concomitantly measured has several major advantages: i) it has the potential for screening a large number of feed resources, for example in breeding programmes for the development of varieties and cultivars of good nutritional value, ii) it could also be of great value in the development of supplementation strategies using locally available conventional and unconventional feed constituents to achieving maximum microbial efficiency in the rumen; iii) it has an important role to play in the study of rumen modulators for increasing efficiency of microbial protein synthesis and decreasing emission of methane, an environmental polluting gas, and iv) it provides a better insight into nutrient-antinutrient and antinutrient-antinutrient interactions, and into the roles of various nutrients (by changing the composition of the incubation medium) with respect to production of fermentative gases, SCFA and microbial mass. The method is also being used increasingly to screen plant-derived rumen modulators. These products have a lower toxicity to animals and humans, and are environmentally friendly. Consequently, they are becoming increasingly popular with consumers.

Further studies are required on: i) the development of simple approaches for identifying the incubation time in the *in vitro* gas system at which the PF (a measure of the proportion of fermented substrate which leads to microbial mass production) is maximum, ii) the effect of nitrogen in the incubation medium on the PF, and iii) the *in vivo* significance of the PF so obtained. The results of the limited experiments conducted so far have shown that simple models employing gas kinetic parameters and the PF are capable of predicting the dry matter intake of roughages and level of emission of methane by ruminants. Experiments also need to be done to test whether, for any given feed, the microbial protein synthesis as derived from digestion kinetic parameters (including PF) *in vitro* is sufficient to explain the observed microbial protein supply to the small intestine *in vivo*. At present, the simplest way of determining the latter parameter is to calculate it from the level of urinary purine derivatives. This validation exercise should be conducted for a wide range of feed constituents and diets which should enable the above mentioned simple technique of measuring gas and microbial mass to be a routine and powerful tool for feed evaluation thus avoiding the need for time-consuming, laborious and expensive feeding studies. Lately, much emphasis has been given to the development of statistical or mathematical models that best fit the gas production profiles and describe the gas evolution with high accuracy. Experiments must be designed to understand the biological significance of the various statistical and functional parameters being calculated using these models, and also to incorporate a measure of microbial mass into these mathematical descriptions.

Research and development efforts are required to establish a feed library for unconventional feedstuffs that includes information on nutritive values in addition to routine composition analysis. In the case of tannin-containing feedstuffs, there is a need to incorporate approach(s) measuring the biological activities of tannins as well as measuring tannin levels by chemical methods.

REFERENCES

Aiple, K.-P., Steingass, H. & Drochner, W. 1996. Prediction of net energy content of raw materials and compound feeds for ruminants by different laboratory methods. *Arch. Anim. Nutr.*, 49: 213–220.

Ben Salem, H., Nefzaoui, A., Ben Salem, L. & Tisserand, J.L. 1999. Different means of administering polyethylene glycol to sheep: effect on the nutritive value of *Acacia cyanophylla* Lindl. Foliage. *Anim. Sci.*, 68: 809–818.

Blümmel, M., Steingass, H., Becker, K. & Soller, H. 1993. Production of SCFA, CO2, CH4 and microbial cells *in vitro*. *Proc. Soc.Nutr. Physiol.* 1: 9.

Blümmel, M., Makkar, H.P.S. & Becker K. 1997. *In vitro* gas production: a technique revisited. *J. Anim. Physiol. Anim. Nutr.*, 77: 24–34.

Blümmel, M., Makkar, H.P.S. & Becker, K. 1998a. The partitioning kinetics of *in vitro* rumen fermentation products in a gas production test, *J. Dairy Sci.. (Suppl. 1), 309.*

Blümmel M., Makkar H.P.S. & Becker K. 1998b. The *in vitro* gas production characteristics of whole roughage versus extracted neutral-detergent fibre and their implications for analysing the fermentation of cell solubles by a differential approach. *In* E.R. Deaville, E. Owen, A.T. Adesogan, C. Rymer, J.A. Huntington & T.L.J. Lawrence, eds. *In* vitro *techniques for measuring nutrient supply to ruminants BSAS* No. 22, pp 85–88, Edingburgh.

Blümmel, M., Aiple, K.-P., Steingass, H. & Becker, K. 1999a. A note on the stoichiometrical relationship of short chain fatty acid production and gas evolution *in vitro* in feedstuffs of widely differing quality. *J. Anim. Physiol. Anim. Nutr.*, 81: 157–167.

Blümmel, M., Mgomezulu, R., Chen, X. B., Makkar H.P.S., Becker K. & Ørskov E.R. 1999b. The modification of in *in vitro* gas production test to detect roughage related differences in *in vivo* microbial protein synthesis as estimated by the excretion of purine derivates. *J. Agric. Sci. (Camb.),* 133: 335–340.

Blümmel, M. & Ørskov, E.R. 1993. Comparison of gas production and nylon bag degradability of roughages in predicting feed intake in cattle. *Anim. Feed Sci. Technol.*, 40: 109–119.

Chen, X.B., Hovell, F.D.D. & Ørskov, E.R. 1990. Excretion of purine derivatives by ruminants: endogenous excretion differences between cattle and sheep. *Br. J. Nutr.*, 63: 121–129.

Chen, Y-K., Pell, A.N., Chase, L.E. & Schofield, P. 1999. Rate and extent of the ethanol-soluble and neutral detergent-insoluble fractions of corn grain. *J. Anim. Sci.*, 77: 3077–3083.

De Boever, J.L., Cottyn, B.G., Buysse, F.X., Wainman, F.W. & Vanacker, J.M. 1986. The use of an enzymatic technique to predict digestibility, metabolisable and net energy of compound feedstuffs for ruminants. *Anim. Feed Sci. Technol.*, 14: 203–214.

Degen, A.A., Becker, K., Makkar, H.P.S. & Borowy, N. 1995. *Acacia saligna* as a fodder tree for desert livestock and the interaction of its tannins with fibre fractions. *J. Sci. Food Agric.*, 68: 65–71.

Dhanoa, M.S. 1988. On the analysis of dacron bag data for low degradability feeds. *Grass and Forage Science*, 43: 441–444.

Dijkstra, J., France, J., Dhanoa, M.S. & Lopez, S. 2000. Simulation of substrate degradation, microbial synthesis and gas production in the gas production technique and *in vivo*. *EAAP Satellite Symposium, Gas production: fermentation kinetics for feed evaluation and to assess microbial activity*, 18-19 August, Wageningen, The Netherlands.

France, J., Dhanoa, M. S., Theodorou, M. K., Lister, S. J., Davies, D. R. & Isac, D. 1993. A model to interpret gas accumulation profiles associated with *in vitro* degradation of ruminal feeds. *J. Theor. Biol.*, 163: 99–111.

France, J., Dijkstra, J., Dhanoa, M. S., Lopez, S. & Bannink, A. 2000. Estimating the extent of degradation of ruminant feeds from a description of their gas production profiles observed *in vitro*: derivation of models and other mathematical considerations. *Brit. J. Nutr.*, 83: 43–150

Getachew, G., Blümmel, M., Makkar H.P.S. & Becker K. 1998a. *In vitro* gas measuring techniques for assessment of nutritional quality of feeds: a review. *Anim. Feed Sci. Technol.*, 72: 261–281.

Getachew, G., Makkar, H.P.S. & Becker, K. 1998b. The *in vitro* gas coupled with ammonia nitrogen measurement for evaluation of nitrogen degradability in low quality roughages using incubation medium of different buffering capacity. *J. Sci. Food Agric.*, 77: 87–95.

Getachew, G., Makkar, H.P.S. & Becker, K. 2000a. Stoichiometric relationship between short chain fatty acid and *in vitro* gas production in presence and absence of polyethylene glycol for tannin containing browses, *EAAP Satellite Symposium, Gas production: fermentation kinetics for feed evaluation and to assess microbial activity*, 18-19 August, Wageningen, The Netherlands.

Getachew, G., Makkar, H.P.S. & Becker, K. 2000b. Effect of polyethylene glycol on *in vitro* degradability of nitrogen and microbial protein synthesis from tannin-rich browse and herbaceous legumes. *Brit. J. Nutr.*, 84: 73–83.

Getachew, G., Makkar, H.P.S. & Becker, K. 2000c. Tannins in tropical browses: effects of tannins on *in vitro* microbial fermentation and microbial protein synthesis in medium containing different amounts of nitrogen. *J. Agric. Food Chem.*, 48: 3581–3588.

Getachew, G., Makkar, H.P.S. & Becker, K. 2000d. Effect of different amounts and method of application of polyethylene glycol on efficiency of microbial protein synthesis in an *in vitro* system containing tannin rich browses, *EAAP Satellite Symposium, Gas production: fermentation kinetics for feed evaluation and to assess microbial activity*, 18-19 August, Wageningen, The Netherlands.

Goering, H. K. & Van Soest, P.J. 1970. Forage fibre analyses (Apparatus, Reagents, Procedures, and Some Applications). *Agric. Handbook* No. 379, ARS-USDA, Washington, DC.

IAEA. 1997. *Estimation of rumen microbial protein production from purine derivatives in urine*. IAEA-TECDOC-945.

IAEA. 1998. Development, standardization and validation of nuclear-related technologies for measuring microbial protein supply in ruminant livestock for improving productivity. Final Research Coordination Meeting, 24–28 August, Vienna, Austria. *Animal Production and Health Newsletter*, 29: 12–15.

Makkar, H.P.S. 1989. Protein precipitation methods for quantitation of tannins: a review. *J. Agric. Food Chem.*, 37: 1197–1202.

Makkar, H.P.S. 2000. *Quantification of tannins in tree foliage- a laboratory manual*; a joint FAO/IAEA working document, Vienna, Austria.

Makkar, H.P.S. & Becker, K. 1999. Purine quantification in digesta from ruminants by spectrophotometic and HPLC methods. *Brit. J. Nutr.*, 81: 107–112.

Makkar, H.P.S., Blümmel, M., Borowy, N.K. & Becker, K. 1993. Gravimetric determination of tannins and their correlations with chemical and protein precipitation methods. *J. Sci. Food Agric.*, 61: 161–165.

Makkar, H.P.S., Blümmel, M. & Becker, K. 1995a. *In vitro* effects of and interactions between tannins and saponins and fate of tannins in the rumen. *J. Sci. Food Agric.*, 69: 481–493.

Makkar, H.P.S., Blümmel, M. & Becker, K. 1995b. Formation of complexes between polyvinyl pyrrolidone and. polyethylene glycol with tannins and their implications in gas production and true digestibility in *in vitro* techniques. *Brit. J. Nutr.*, 73: 897–913.

Makkar, H.P.S., Borowy, N.K., Becker, K. &d Degen, A. 1995c. Some problems in fiber determination of a tannin-rich forage (*Acacia saligna* leaves) and their implications in *in vivo* studies. *Anim. Feed Sci. Technol.*, 55: 67–76.

Makkar, H.P.S., Blümmel, M. & Becker, K. 1997a. *In vitro* rumen apparent and true digestibility of tannin-rich forages. *Anim. Feed Sci. Technol.*, 67: 245–251.

Makkar, H.P.S., Becker, K. & Younan, M. 1997b. Nutritional implications of bound proanthocyanidins. *Proc. XVIII International Grassland Congress*, June 8–12, Winnipeg, Canada.

Makkar H. P. S., Blümmel M. & Becker K. 1998a. Application of an *in vitro* gas method to understand the effects of natural plant products on availability and partitioning of nutrients. *In* E.R. Deaville, E. Owen, A.T. Adesogan, C. Rymer, J.A. Huntington & T.L.J. Lawrence, eds. In vitro *techniques for measuring nutrient supply to ruminants*, pp 147–150, Edingburgh: BSAS No 22.

Makkar, H.P.S., Sen, S., Blümmel, M. & Becker, K. 1998b. Effects of fractions containing saponins from *Yucca schidigera, Quillaja saponaria* and *Acacia auriculoformis* on rumen fermentation. *J. Agric. Food Chem.*, 46: 4324–4328.

McDonald, I. 1981. A revised model for the estimation of protein degradability in the rumen. *J. Agric. Sci. (Camb.)*, 96: 251–252.

Mehrez, A.Z. & Ørskov, E.R. 1977. A study of artificial fibre bag technique for determining the digestibility of feeds in the rumen. *J. Agric. Sci. (Camb.)*, 88: 645–650.

Menke, K. H., Raab, L., Salewski, A., Steingass, H., Fritz, D. & Schneider, W. 1979. The estimation of the digestibility and metabolizable energy content of ruminant feedstuffs from the gas production when they are incubated with rumen liquor *in vitro*. *J. Agric. Sci. (Camb.)*, 92: 217–222.

Raab, L., Cafantaris, B., Jilg, T. & Menke, K.H. 1983. Rumen protein degradation and biosynthesis. 1. A new method for determination of protein degradation in rumen fluid *in vitro*. *Brit. J. Nutr.*, 50: 569–582.

Rymer, C. & Givens, D.I. 1999. The use of the *in vitro* gas production technique to investigate the effect of substrate on the partitioning between microbial biomass production and the yield of fermentation products. Proc. *Br. Soc. Anim. Sci.*, pp 36.

Sampath, K.T., Wood, C.D. & Prasad, C.S. 1995. Effect of urea and by-products on the *in vitro* fermentation of untreated and urea treated finger millet (*Eleusine coracana*) straw. *J. Sci. Food Agric.*, 67: 323–328.

Tilley, J.M. & Terry, R.A. 1963. A two stage technique for the *in vitro* digestion of forage crops. *J. Brit. Grassl. Soc.*, 18: 104–111.

Van Soest, P.J. 1994. *Nutritional ecology of ruminants*. 2nd edition. Cornell University Press.

Van Soest, P.J., Robertson, J.B. & Lewis, B. 1991. Methods for dietary fiber, neutral detergent fiber, and nonstarch polysaccharides in relation to animal nutrition. *J. Dairy Sci.*, 74: 3583–3597.

Wolin, M.J. 1960. A theoretical rumen fermentation balance. *J. Dairy Sci.*, 43: 1452–1459.

Microbiology of animal feeds

J.P.F. D'Mello
Formerly of the Scottish Agricultural College (SAC)
Edinburgh, United Kingdom

A wide range of microbes occurs naturally on, or as contaminants of forages, cereal grains, oilseed by-products and compound feeds. Beneficial effects can result from the occurrence of the lactic acid bacteria during the fermentation of forages during the process of ensilage. These bacteria favour the production of lactic acid, which helps in reducing the pH to around 4.0, thus preserving the forage for winter-feeding to ruminants. Lactic acid bacteria and yeast cultures have also been attributed with beneficial properties as feed probiotics for reducing scouring and increasing growth performance in farm animals. Animal feeds may become contaminated with harmful bacteria such as *Salmonella*, *Listeria* and *E. coli*. In the case of the latter organism, faecal sources and slurry have been identified as the primary routes of contamination both in pastures and compound feeds. Cereal grains and oilseed by-products are regularly contaminated with fungi occurring as plant pathogens or developing during storage. Major adverse effects arise in farm animals due to the production of mycotoxins by certain species and strains of these fungi. The use of fungicides and preservatives represents potential methods for reducing the prevalence of deleterious fungi. Extensive legislation exists for the control of a number of feed contaminants, particularly mycotoxins. These are reviewed together with current bovine spongiform encephalopathy (BSE) regulations.

1. INTRODUCTION

The microbiology of animal feeds emerged as an important issue in the wake of the *Salmonella*, *E. coli* O157, *Campylobacter* and BSE crises in the European Union (EU) and elsewhere and the momentum has been maintained by the foot and mouth epidemic of 2001/2 in the United Kingdom (UK). Fungal contamination of animal feeds is a regular occurrence on a worldwide scale and detrimental effects have been observed in all classes of farm animals due to the production of mycotoxins by certain species and strains of moulds. There are few grounds for complacency as regards animal feed safety and vigilance should now be the watchword for all those involved in the livestock industry. Nevertheless, it should be recognized that successful silage production depends upon the promotion of fermentation brought about by beneficial bacteria. This review will focus on several aspects of feed microbiology, including silage fermentation, fungal and bacterial contaminants, probiotics and legislation. Where relevant, comments on BSE legislation will be included to provide a more integrated picture of current developments and future strategy.

2. SILAGE MICROBIOLOGY

Successful preservation of high-moisture forage and other crops depends upon the controlling the activities of microbes, particularly bacteria. Conditions favouring optimal preservation include the rapid imposition and subsequent maintenance of anaerobic conditions at all stages in the process of ensilage. Under these conditions, lactic acid bacteria proliferate, using endogenous plant sugars to produce sufficient quantities of the acid to depress pH to around 4, the optimum value for successful preservation. Wet forages are difficult to preserve in this way and often provide ideal conditions for the growth of undesirable bacteria such as the Clostridia. The features of both types of bacteria are summarized in Table 1.

Species

The lactic acid bacteria (LAB) are of two types. The homofermentative species include *Lactobacillus plantarum*, *Pediococcus pentosaceus* and *Enterococcus faecalis*. The heterofermentative species comprise *Lactobacillus brevis* and *Leuconostoc mesenteroides*. The homofermentative group are more efficient in converting forage sugars into lactic acid (Table 1). Of these species *L. plantarum* is

TABLE 1
Microbiology of ensilage[1]

Organisms	Conditions required	Major products/effects
Lactic acid bacteria (LAB)	Anaerobic; wilting of crop is desirable; crop should be chopped for rapid establishment of LAB.	Homofermentative pathway: lactic acid and some acetic acid. Heterofermentative pathway: lactic acid, ethanol, mannitol, acetic acid and CO2.
Clostridia	Anaerobic; wet forage.	Saccharolytic species: butyric acid, CO2 and H2. Proteolytic species: butyric acid, acetic acid, amines, CO2 and NH3.
Enterobacteria	Anaerobic; optimum pH 7.0; active in early stages of fermentation.	Acetic acid, ethanol, CO2 H2 and NH3.
Listeria	Aerobic; pH above 5.5; growth possible at low temperatures and in high-dry matter silages	Listeriosis, especially in sheep.
Fungi	Aerobic; active on surface layers of silage.	Spores and mycotoxins.

[1]Compiled from McDonald *et al.* (1991).

generally considered to be the most competitive, rapidly producing large amounts of lactic acid in freshly ensiled forage. LAB are essentially non-proteolytic organisms and are therefore considered to be beneficial in that they assist in conserving labile proteins and free amino acids in forage.

Two groups of Clostridia are also recognized. The saccharolytic group includes *Clostridium butyricum* and *C. tyrobutyricum*. These species ferment residual sugars as well as lactic acid to butyric acid, causing a rise in pH. The proteolytic group includes *C. bifermentans* and *C. sporogenes*. These bacteria ferment amino acids to a number of different products (Table 1). The Clostridia thrive under high-moisture conditions and consequently the ensilage of wet forages is fraught with problems associated with adequate preservation and satisfactory voluntary intakes and performance of animals fed on such silages.

The Enterobacteria include *Escherichia coli* and *Erwinia herbicola* and are also considered to be undesirable in that they compete with

LAB for plant sugars fermenting them to acetic acid, ethanol, CO_2 and H_2. They are also capable of catabolising amino acids to NH_3 (Table 1).

Listeria monocytogenes is widely distributed in nature and also occurs in silage, particularly big bale silage. The increased incidence of listeriosis in sheep and cattle has been partly linked with the introduction of big bale silage. The relatively low density and limited fermentation characteristics of big bale silage and susceptibility of the bags to damage all favour the growth of *L. monocytogenes*. This organism is of particular significance because of it potential to contaminate animal products destined for human consumption.

Fungi occur in silages as yeasts and moulds. The yeasts include species of *Candida*, *Saccharomyces* and *Torulopsis*. The moulds associated with silages include various species of *Aspergillus*, *Penicillium* and *Fusarium*. The occurrence of these fungi is of particular concern due to their potential to produce harmful mycotoxins.

Modulation of microbial activity

The activities of bacteria may be controlled to facilitate preservation of ensiled forages. Wilting the crop before ensiling is a common method of restricting fermentation, allowing the growth of LAB but inhibiting the activities of undesirable organisms such as Clostridia and Enterobacteria. An added benefit of wilting is that effluent production is reduced, thus minimising environmental pollution from this source.

Additives may be used to control or stimulate fermentation by bacteria in ensiled forages. Formic acid is widely used to restrict fermentation by artificially reducing pH to below 4.0. Commercial products are available containing, for example, 85 percent formic acid which is applied during the harvesting of the forage. However, even at the recommended rates of application, the bacteria are still able to ferment sugars to lactic acid. Stimulation of silage bacteria may be brought about in two ways. Addition of molasses to the forage prior to ensiling ensures sufficient substrate for LAB thus causing the desired reduction in pH values. More recently, however, it has been suggested that a limiting factor is the relatively low concentrations of LAB in forage crops at harvest. Consequently, a number of commercial inoculants containing freeze-dried cultures of homofermentative LAB have become available for application to harvested forage. Efficacy of this procedure depends upon the inoculation rate as well as the presence of adequate concentrations of sugars.

3. FUNGAL CONTAMINATION OF CONCENTRATES AND FORAGES

Fungal contamination of cereal grains, oil-seed meals and forages continues to represent a major animal health risk throughout the world and particularly in the humid tropics. The risks arise primarily from the ability of particular species and strains of fungi to produce harmful compounds known as mycotoxins. These substances arise from the secondary metabolism of fungi in response to a wide range of genetic and environmental factors.

Mycotoxin contamination of forages and cereals frequently occurs in the field following infection of plants with specific pathogenic fungi or with symbiotic endophytes. In addition, contamination may occur during processing and storage of harvested products and feed whenever environmental conditions are appropriate for spoilage fungi. Moisture content and ambient temperature are key factors affecting fungal colonisation of and mycotoxin production in concentrates and compound feeds.

Inhalation of fungal spores or consumption of mycelia may cause conditions collectively known as `mycoses'. An example of such a condition is mycotic abortion arising from the systemic transmission and delivery of fungal material to the placenta.

Classification

It is conventional to subdivide toxigenic fungi into `field' (or plant pathogenic) and `storage' (or saprophytic/spoilage) organisms. *Claviceps*, *Neotyphodium Fusarium* and *Alternaria* are classical representatives of field fungi while *Aspergillus* and *Penicillium* exemplify storage organisms. Mycotoxigenic species may be further distinguished on the basis of geographical prevalence, reflecting specific environmental requirements for growth and secondary metabolism. Thus, *Aspergillus flavus*, *Aspergillus parasiticus* and *Aspergillus ochraceus* readily proliferate under warm, humid conditions whereas *P. expansum* and *P. verrucosum* are essentially temperate fungi. Consequently, the *Aspergillus* mycotoxins predominate in plant products emanating from the tropics and other warm regions, while the *Penicillium* mycotoxins occur widely in temperate foods, particularly cereal grains. *Fusarium* fungi are more ubiquitous, but even this genus contains toxigenic species which are almost exclusively associated with cereals from tropical and sub-tropical countries. The distribution of the important fungi in concentrates and forages is shown in Table 2 which also provides a list of the major mycotoxins.

The *Aspergillus* genus dominates all other fungi in respect of mycotoxin production in cereals and oilseeds. For example, Dhand *et al.* (1998) observed that *Aspergillus* was the most significant genus in dairy and other feeds in the tropics. Three species are responsible for virtually all mycotoxin production by this genus: *Aspergillus flavus*, *Aspergillus parasiticus* and *Aspergillus ochraceus*. *A. flavus*, and *A. parasiticus* synthesize the aflatoxins, while *A. ochraceus* produces the ochratoxins.

Aspergillus species

The aflatoxins include aflatoxin B1, B2, G1 and G2 (AFB1, AFB2, AFG1 and AFG2, respectively). In addition, aflatoxin M1 (AFM1) may well occur in the milk of dairy cows consuming AFB1-contaminated feeds. The aflatoxigenic Aspergilli are generally regarded as storage fungi, proliferating under conditions of relatively high moisture / humidity and temperature. Aflatoxin contamination is, therefore, almost exclusively confined to tropical feeds such oilseed by-products derived from groundnuts, cottonseed and palm kernel. Aflatoxin contamination of maize is also an important problem in warm humid regions where *A. flavus* may infect the crop prior to harvest and remain viable during storage.

TABLE 2
Toxigenic fungi of concentrates and forages[1]

Fungi	Occurrence	Mycotoxins
Aspergillus flavus; A. parasiticus	Peanut meal, cottonseed cake, palm kernel cake, maize, compound feeds	Aflatoxins
A. flavus	Oilseed meals, compound feeds	cyclopiazonic acid
A. ochraceus; Penicillium viridicatum; P. cyclopium	Barley and wheat grains	Ochratoxin A
P. citrinum; P. expansum	Cereal grains	Citrinin
P. citreo-viride	Cereal grains	Citreoviridin
Fusarium culmorum; F. graminearum	Cereal grains	Deoxynivalenol
F. sporotrichioides; F.poae	Cereal grains	T-2 toxin
F. sporotrichioides; F. graminearum; F. poae	Cereal grains	Diacetoxyscirpenol
F. culmorum; F. graminearum; F. sporotrichioides	Cereal grains	Zearalenone
F. moniliforme	Maize kernels	Fumonisins; moniliformin; fusaric acid
Neotyphodium coenophialum	Grasses	Ergopeptine alkaloids
N. lolii	Grasses	Lolitrem alkaloids
Claviceps purpurea	Cereal grains	Ergot alkaloids
Phomopsis leptostromiformis	Lupin stubble	Phomopsins
Pithomyces chartarum	Pastures	Sporidesmin A

[1]Adapted from D'Mello (2000).

As stated above, *A. ochraceus* produces the ochratoxins. However it shares this property with at least two *Penicillium* species. Ochratoxin A (OTA) and ochratoxin B are two forms that occur naturally as contaminants, with OTA being more ubiquitous, occurring predominantly in cereal grains and in the tissues of animals given contaminated feed. Another mycotoxin, citrinin, often co-occurs with OTA, particularly in wheat.

Fusarium species

The important *Fusarium* fungi of cereals include: *F. graminearum*, *F. culmorum*, *F. sporotrichioides*, *F. poae* and *F. moniliforme* (Table 2). These species produce a wide range of mycotoxins. Of particular significance in animal health are the trichothecenes, zearalenone and the fumonisins. The trichothecenes are sub-divided into four basic groups, with types A and B representing the most important members. Type A trichothecenes include T-2 toxin, HT-2 toxin, neosolaniol and diacetoxyscirpenol. Type B trichothecenes include deoxynivalenol (also known as vomitoxin), nivalenol and fusarenon-X. The production of the two types of trichothecenes is characteristic for a particular *Fusarium* species. However, a common feature of the secondary metabolism of these fungi is their ability to synthesize zearalenone which, consequently, occurs as a co-contaminant with certain trichothecenes. The fumonisins are synthesized by a distinct group of *Fusarium* species (Table 2). Three members of this group (fumonisin B1, B2 and B3) often occur together in maize.

Virtually all the toxigenic species of *Fusarium* described above are also major pathogens of cereal plants, causing diseases such as head blight in wheat and barley and ear rot in maize. Harvested grain from diseased crops is, therefore, likely to be contaminated with the appropriate mycotoxins, and this is supported by ample evidence.

Neotyphodium coenophialum

The endophytic fungus *Neotyphodium coenophialum* occurs in close association with perennial tall fescue, while another related fungus, *N. lolii*, may be present in perennial ryegrass (D'Mello, 2000). Ergopeptine alkaloids, mainly ergovaline, occur in *N. coenophialum*-infected tall fescue, while the indole isoprenoid lolitrem alkaloids, particularly lolitrem B are found in *N. lolii*-infected perennial ryegrass. The ergopeptine alkaloids reduce growth, reproductive performance and milk production in cattle, while the lolitrem compounds induce neurological effects in ruminants.

Phomopsis leptostromiformis

In Australia, lupin stubble is valued as fodder for sheep, but infection with the fungus *Phomopsis leptostromiformis* is a major limiting factor due to toxicity arising from the production phomopsins by the fungus. Mature or senescing parts of the plant, including stems, pods and seeds are particularly prone to infection. Phomopsin A is considered to be the primary toxin, causing effects such as ill-thrift, liver damage, photosensitization and reduced reproductive performance in sheep (D'Mello and Macdonald, 1998).

Pithomyces chartarum

Pithomyces chartarum is a ubiquitous saprophyte of pastures with the capacity to synthesize sporidesmin A, a compound causing facial eczema and liver damage in sheep.

Use of organic acids

Fungal contamination of feeds, particularly high-moisture cereal grains may be prevented by the use of organic acids prior to storage. Propionic acid has emerged as an effective agent for this purpose. However, mixtures of organic acids are now commercially available. The efficacy of BASF mixtures has recently been undertaken in two studies conducted at the Scottish Agricultural College (SAC) and published by Blanchard *et al.* (2001). The preservatives tested included Lupro-Mix® NC (containing a mixture of 38 percent propionic acid, 34 percent formic acid, 8 percent ammonia and 20 percent water) and Lupro-Grain® (a mixture of 92 percent, propionic acid, 4 percent ammonia and 4 percent propandiol). The efficacy was compared with propionic acid at similar levels of addition (0.10, 0.25 and 0.50 percent). In the first investigation, the three preservatives were tested against 6 species of storage fungi using a purified medium (potato dextrose agar, PDA) for culture of these moulds. The minimum inhibition concentration (MIC) was then determined for the test products. As shown in Table 3, propionic acid emerged as the most effective and consistent preservative, with Lupro-Mix® being more effective than Lupro-Grain® in controlling *Penicillium expansum*. In contrast, Lupro-Grain® was more effective than Lupro-Mix® for the suppression of *Aspergillus flavus* and *A. parasiticus*, while both were equally efficient against *A. ochraceus* and *P. verrucosum* and *P. roqùefortii*. *P. verrucosum* was more effectively controlled by the four test compounds than the other storage fungi under the conditions of this investigation.

TABLE 3
Minimum inhibition concentrations of three test products against six storage fungi inoculated on purified substrate (potato dextrose agar) [1]

Product	*A. flavus*	*A. parasiticus*	*A. ochraceus*	*P. roquefortii*	*P. expansum*	*P. verrucosum*
Propionic acid	0.25	0.25	0.25	0.25	0.25	0.10
Lupro-Mix® NC	0.50	0.50	0.25	0.50	0.25	0.25
Lupro-Grain®	0.25	0.25	0.25	0.50	0.50	0.25

[1] Taken from Blanchard *et al.* (2001)

In the second trial, moist sterilized barley grain was inoculated with A. parasiticus and treated with 0.10, 0.25 and 0.50 percent of the three test products. In contrast to the findings of the first study, Lupro-Grain® emerged as the most effective product with the lowest MIC value (Table 4). Substrate type influenced the efficacy of test compounds against *A. parasiticus*. Discrepancies may have been caused by differences in water activity (i.e. moisture content) or in buffering capacity of the two substrates. Clearly much more work needs to be done to fully assess the efficacy of these preservatives in other cereals and in compound feeds.

4. *SALMONELLA* SPECIES

Many species of *Salmonella* have been implicated in diseases of farm animals. Of these, *S. typhimurium* is universally distributed while *S. enteriditis* has emerged as a regular pathogen of poultry and contaminant of eggs and chicken meat. Animal feeds are thought to be an important source of these bacteria. Meat and bone meal and fisheal

TABLE 4
Minimum inhibition concentration of three test products against *A. parasiticus* inoculated on to barley grain[1,2]

Product	Minimum inhibition concentration
Propionic acid	0.25
Lupro-Mix® NC	0.25
Lupro-Grain®	0.10

[1] Taken from Blanchard *et al.* (2001).
[2] 100g sterilised barley grain used with moisture content adjusted to 28 percent.

are frequently contaminated with *Salmonella*. Intensive pasture utilisation provides an additional source through contamination of faeces from infected animals. Furthermore, the practice of spreading cattle slurry on to pastures in conventional and organic farms is another potentially significant source of infection.

In many parts of the world poultry manure is used as a feed for ruminants. For example in the USA, two poultry waste products are available for such use: dried poultry waste and dried poultry litter (Jeffrey *et al.*, 1998). Dried poultry waste represents undiluted excreta generally derived from caged layer flocks, whereas dried poultry litter is a mixture of excreta and litter. These products are heated to reduce bacterial contamination but are, nevertheless, not sterile. It is reassuring to note that in the trail of Jeffrey *et al.* (1998) that all samples of processed poultry litter collected from 13 dairy farms were totally free of *Salmonella* contamination even though virtually all samples contained Enterobacteriacea, non-glucose fermenting Gram-negative and Gram-positive bacteria.

5. *ESCHERICHIA COLII* INCLUDING *E.COLI* O157

It is widely recognized that cattle feeds contain *E. coli* through contamination with faeces. There is particular concern over the occurrence of *E. coli* O157 since this form has been definitively linked with specific outbreaks of illness in humans.

The application of slurry on to pastures means that there is potential for the transfer of faecal *E. coli* to grazing animals, a practice that has caused some disquiet among those concerned with food safety. Hancock *et al.* (1998) maintain that it is important to distinguish the notion of 'reservoir' from that of an incidental host as regards the source of *E. coli* O157. They were not able to support the thesis that cattle are the reservoir species of this form of the bacterium. The Pennington Group (1997) also accepted that impact of spreading slurry on land still requires evaluation but proceeded to recommend the introduction of precautionary measures with this practice. Thus, the contamination of pasture grass with *E. coli* O157 from slurry remains a contentious issue.

Other forms of *E. coli* occur extensively in cattle feeds. Thus, Lynn *et al.* (1998) reported that just over 30 percent of cattle feed samples collected from 13 dairies and four feed mills in USA were found to be contaminated with *E. coli*, although *E. coli* O157 was not detected in any of these samples. In five dairies, concentrations of *E. coli* exceeded 1000 colony forming units/g feed. These authors suggested that attention should focus on the replication of *E. coli* in moist feeds and duration of storage in feed bunks. Other studies in the USA also demonstrated the occurrence of *E. coli* (non-O157) in cattle feeds. Jeffrey *et al.* (1998) reported that 13 non-O157 *E. coli* bacteria were isolated from 52 samples of dried poultry litter destined for dairy feeds.

7. FOOT AND MOUTH VIRUS

Foot and mouth disease has been a major problem in the UK during 2001/2, with devastating effects on the livestock industry. The disease is spread by inhalation and by ingestion of contaminated materials. The outbreak in the UK has been tentatively attributed to the feeding of catering waste to pigs. While this association may never be confirmed, it is clear that in the intensive rearing of animals due recognition should be given to the need to correctly process animal feeds. However, in the EU, the feeding of catering waste containing meat products is banned under a recent directive. Thus, in theory at least, this source of the foot and mouth virus should be excluded from livestock farms.

8. BOVINE SPONGIFORM ENCEPHALOPATHY (BSE)

The prion proteins of processed animal proteins have recently emerged as important feed contaminants implicated in the development of BSE in cattle. Prion proteins are normal animal tissue components with the capacity to transform into agents causing fatal neurological syndromes in a wide range of species. The significance of prions has been highlighted following the emergence of bovine spongiform encephalopathy (BSE) as a major disease of cattle in the United Kingdom. The onset of this disease was attributed to the feeding of cattle with meat-and-bone meal prepared from carcasses of scrapie-infected sheep. The latter disease is also caused by prion proteins as is the human equivalent, new variant Creutzfeldt - Jakob disease (vCJD). The incidence of vCJD in humans has been linked with the consumption of BSE-contaminated beef. It is this association that has led to extensive and stringent legislation in the European Union concerning the use of specified animal products in livestock feeding.

9. PROBIOTICS

A number of microbial products added to the feed may exert beneficial effects on livestock health and production. These products are not attributed with specific nutritional roles and are given the term `probiotics'. Lactobacillus acidophilus, for example, appears to be endowed with the ability to reduce scouring and increase liveweight gain in calves, but the effects are not consistent between trials. Yeast cultures may also be used as probiotics (Wallace and Newbold, 1992). It should be stressed that the mechanisms of action of probiotics have yet to be elucidated and there is a need for critical cost-benefit analysis of the various products as they become available for commercial use.

10. LEGISLATION

It might be expected that bacterial contamination of feeds and forages should be comprehensively regulated by statutory measures. However, in the United Kingdom, regulations only exist for *Salmonella* contamination. For example, under UK regulations positive identification of *Salmonella* in feeds must be reported to a `veterinary officer of the Minister' (HMSO, 1989). Current evidence regarding *E. coli* reservoirs is not sufficiently convincing to warrant prohibition of the spreading of slurry or manure on to pasture. It is difficult to see how faecal sources of this bacterium in ruminant feeds and pastures can be reduced by the introduction of regulations. Any introduction of such legislation will require prior evaluation of the critical control points involved in the initial contamination and subsequent spread of *E. coli* on farms.

The acute toxicity of fungal secondary compounds and in particular the carcinogenicity of certain mycotoxins has led to the establishment of regulatory and advisory directives for primary feed ingredients and complete rations. Table 5 is designed to illustrate the type of directives in place in the United Kingdom, European Union and in North America. As can be seen, extensive regulations exist for the aflatoxins, but for ochratoxin A and deoxynivalenol only advisory directives have been proposed, unsupported by legislative measures. Of particular concern is the absence of statutory regulations for the carcinogenic fumonisins.

Animal feeds have been subject to additional stringent controls in the aftermath of the BSE crisis in the United Kingdom and elsewhere in Europe. In the UK, it has been illegal to feed ruminants with any form of mammalian protein since November 1994. The feeding of mammalian meat-and-bone meal (mMBM) to any farmed livestock has been prohibited since April 1996. The latter ban was introduced to avoid inadvertent feeding of mMBM to ruminants. Surveillance results of animal feeds indicate widespread compliance with these regulations, with 99.7 percent of feeds found to be negative for prohibited proteins. EU-wide feed legislation is also in place regarding BSE control of feeds. Thus, the Regulation (EC) No. 1774/2002 of the European Parliament and of the Council of 3 October 2002 requires Member States that only materials derived from animal declared fit for human consumption following veterinary inspection may be used for the production of feeds. It imposes the exclusion of dead animals and other condemned materials from the feed chain, the complete separation during collection, transport, storage, handling and processing of animal waste not intended for animal feed or human food and the complete separation of plants dedicated to feed production from plants processing other animal waste destined for destruction. It also bans intra-species recycling and it sets out clear rules on what must and may be done with the excluded animal materials, imposing strict identification and traceability system requiring certain products such as meat and bone meal and fats destined for destruction to be permanently marked to avoid possible

TABLE 5

Examples of worldwide guidelines and regulations for fungal secondary compounds (mycotoxins) [1]

Countries	Secondary compounds	Commodities	Maximum levels	Status
European Union	Aflatoxin B1 (_g/kg)	Straight feedingstuffs except:	50	Statutory
		groundnut, copra, palm kernel, cottonseed, babassu, maize and products derived from the processing thereof	20	
		Complete feedingstuffs for cattle, sheep and goats (with the exception of complete feedingstuffs for calves, lambs and kids)	50	
		Complete feedingstuffs for pigs and poultry (except those for young animals)	20	
		Other complete feedingstuffs	10	
	Aflatoxin M1 (mg/kg)	Milk/milk products	0.5	Statutory
	Ochratoxin A (_g/kg)	Cereals and cereal products	4	Advisory
	Ergot	All feedingstuffs containing unground cereals	1000	Statutory
Worldwide	Fumonisins	Not regulated	-	-
USA/ Canada	Deoxy-nivalenol (mg/kg)	Animal feed	5-10	Advisory

[1] Adapted from D'Mello (2000).

fraud and risk of diversion of unauthorized products into food and feed. The new Regulation requires the complete disposal, by incineration or landfill after appropriate heat treatment, of Category 1 materials [i.e., animal by-products presenting highest risk such as transmissible spongiform encephalopathies (TSEs) or scrapie]. Category 2 materials (including animal by-products presenting a risk of contamination with other animal diseases) may be recycled for uses other than feed after appropriate treatment (e.g. biogas, composting, oleo-chemical products, etc). Finally, only category 3 materials (i.e., by-products derived from healthy animals slaughtered for human consumption) may be used in the production of feed following appropriate treatment in approved processing plants.

REFERENCES

Blanchard, P.J., D'Mello, J.P.F., Macdonald, A.M.C., Cattan, S. & Roser, U. 2001. Minimum inhibition concentrations for propionic acid and organic acid mixtures against storage fungi. *World Mycotoxin Forum* pp. 62–62. Proceedings of a conference held on 14-15 May 2001 in the Netherlands.

Dhand, N.K., Joshi, D.V. & Jand, S.K. 1998. Fungal contaminants of dairy feed and their toxigenicity. *Indian Journal of Animal Sciences*, 68: 1095–1096.

D'Mello, J.P.F. 2000. Anti-nutritional factors and mycotoxins. *In* J.P.F. D'Mello, ed. *Farm animal metabolism and nutrition*, pp. 383-403.CABI Publishing, Wallingford.

D'Mello, J.P.F. 2002. Mycotoxins in plant products. *In* J.P.F. D'Mello, ed. *Food Safety.* CABI Publishing, Wallingford, (in press).

D'Mello, J.P.F. & Macdonald, A.M.C. 1998. Fungal toxins as disease elicitors. *In* J. Rose, ed. *Environmental toxicology: current developments*, pp. 253-289. Gordon and Breach Science Publishers, Amsterdam,

Hancock, D.D., Besser, T.E., Rice, D.H., Ebel, E.D., Herriott, D.E. & Carpenter, L.V. 1998. Multiple sources of *Escherichia coli* O157 in feedlots and dairy farms in the Northwestern USA. *Preventive Veterinary Medicine*, 35: 11–19.

HMSO. 1989. Statutory instruments. *The zoonoses order* 1989. Her Majesty's Stationery Office, London.

Jeffrey, J.S., Kirk, J.H., Atwill, E.R. & Cullor, J.S. 1998. Prevalence of selected microbial pathogens in processed poultry waste used as dairy cattle feed. *Poultry Science*, 77: 808–811.

Lynn, T.V., Hancock, D.D., Besser, T.E., Harrison, J.H., Rice, D.H., Stewart, N.T. & Rowan, L.L. 1998. The occurrence and replication of *Escherichia coli* in cattle feeds. *Journal of Dairy Science*, 81: 1102–1108.

McDonald, P., Henderson, A.R. & Heron, S.J.E. 1991. *The biochemistry of silage*. Chalcombe Publications, Marlow. 340 pp.

McGee, P., Bolton, D.J., Sheridan, J.J., Earley, B. & Leonard, N. 2001. The survival of *Esherichia coli* O157:H7 in slurry from cattle fed different diets. *Letters in Applied Microbiology,* 32: 152–155.

Pennington Group. 1997. *Report on the circumstances leading to the 1996 outbreak with* E. coli *O157 in central Scotland, the implications for food safety and the lessons to be learned*, pp. 38–39. The Stationery Office, Edinburgh.

Wallace, R.J &Newbold, C.J. 1992. Probiotics for ruminants. *In* R. Fuller, ed. *Probiotics. The scientific basis,* pp. 317–353. Chapman & Hall, London,

McGee, P., Bolton, D.J., Sheridan, J.J., Earley, B. & Leonard, N. 2001 The survival of *Escherichia coli* O157:H7 in slurry from cattle fed different diets. *Letters in Applied Microbiology* 32:152–155.

Pennington Group. 1997. *Report on the circumstances leading to the 1996 outbreak of E. coli O157 in central Scotland, the implications for food safety and the lessons to be learned*, pp. 1–49. The Stationery Office, Edinburgh.

Wallace, R.J. & Newbold, C.J. 1992. Probiotics for ruminants. *In* R. Fuller, ed. *Probiotics. The scientific basis*, pp. 317–353. Chapman & Hall, London.

Contaminants and toxins in animal feeds

J.P.F. D'Mello
Scottish Agricultural College (SAC) in Edinburgh
United Kingdom

This work was partly supported by funding from the Scottish Executive Rural Affairs Department (Project 627012).

In global terms, animal feeds and forages contain a wide range of contaminants and toxins arising from anthropogenic and natural sources. In this article, the distribution of heavy metals, radionuclides, mycotoxins, plant toxins, antibiotics and microbial pathogens in cereals, complete feeds and forages is reviewed. The impacts on farm livestock productivity and on the safety of resulting edible products are also considered. Evidence is provided to demonstrate that feeds contain a variety of substances as co-contaminants and that there are regional differences in the nature of the compounds involved. It is concluded that the options for remedial action are limited. Furthermore, although many developing countries lack appropriate legislation, change in this respect is inevitable as regulatory controls for feeds imported into Europe and America are strengthened.

1. INTRODUCTION

Animal feeds are routinely subject to contamination from diverse sources, including environmental pollution and activities of insects and microbes. Animal feeds may also contain endogenous toxins arising principally from specific primary and secondary substances produced by fodder plants. Thus, feed toxins include compounds of both plant and microbial origin. Although these toxins are often considered separately, because of their different origins, they share several common underlying features. Thus, particular compounds within both plant and microbial toxins may exert antinutritional effects or reduce reproductive performance in farm animals. Furthermore, the combined effects may be the result of additive or synergistic interactions between the two groups of compounds. The extent and impact of these interactions in practical livestock feeding remain to be quantified. Feed contaminants and toxins occur on a global scale but there are distinct geographical differences in the relative impact of individual compounds. The term "feed" is generally used in its widest context to include compound blends of straight ingredients as well as forages. With such a broad perspective, it is necessary and more instructive to introduce some focus. Consequently, this article is limited to a review of those contaminants and toxins that represent significant risks to farm livestock. Feed contamination arising from insect fragments and excreta will not be addressed, but the role of such vectors in the transmission of fungal spores and hyphae should not be ignored. Legal control of certain feed contaminants and toxins is in place and operating within a continually evolving framework; the salient issues will be briefly reviewed here.

2. ENVIRONMENTAL CONTAMINANTS

A wide range of organic and inorganic compounds may occur in feedstuffs, including pesticides, industrial pollutants, radionuclides and heavy metals. Pesticides that may contaminate feeds originate from most of the major groups, including organochlorine, organophosphate and pyrethroid compounds (van Barneveld, 1999). A recent survey indicated that 21 percent of feeds in the United Kingdom contain pesticide residues. Pirimiphos-methyl, an insecticide used in grain stores, was detected with the highest frequency. Although pesticides are potentially toxic to farm livestock, the primary focus of concern centres on residues in animal products destined for human consumption. Dioxins and polychlorinated biphenyls (PCBs) are examples of industrial pollutants that may contaminate feeds, particularly herbage. Cows grazing pastures that are close to industrial areas produce milk with higher dioxin content than cows from rural farms. In 1999, dioxin-contaminated animal fat was inadvertently added to animal feeds destined for Belgian, French and Netherlands farms. Unacceptable levels of dioxins were found in meat products and eggs from these farms.

Human health considerations are also paramount in the monitoring of radionuclide pollution. Following the Chernobyl accident in 1986, caesium-134 and caesium-137 were released, causing widespread contamination of pastures and conserved forages. As a consequence, milk and sheep carcasses became contaminated and restrictions were imposed on the movement and slaughter of sheep (MAFF, 1994).

Contamination of feeds and herbage with cadmium may occur as a result of applying certain types of fertilizers to crops and pastures. On the other hand, lead contamination arises from industrial and urban pollution, while mercury in feeds arises from the use of fishmeal.

3. BACTERIAL CONTAMINANTS

There is currently considerable interest in the occurrence of *Escherichia coli* in animal feeds following the association of the O157 type of these bacteria with human illness. In a recent United States study (Lynn *et al.*, 1998), 30 percent of cattle feed samples obtained from commercial sources and farms contained *E. coli*, although none of the tests for *E. coli* O157 were positive. Replication of faecal *E. coli*, including the O157 type, was demonstrated in a variety of feeds under conditions likely to occur on cattle farms in the summer months. Since faecal contamination of feeds is widespread on farms, it is an important route for exposure of cattle to *E. coli* and other organisms. The potential for exposure to bacteria also exists when poultry litters are fed to cattle (in California, for example, two such poultry waste products are commercially available for use as cattle feed). However, providing the products are adequately heat-processed prior to distribution, the risks of contamination with *E. coli*, *Salmonella* spp. and *Campylobacter* spp. are likely to be minimized or even eliminated (Jeffrey *et al.*, 1998). Nevertheless, it is worth noting that *S. enterica* commonly occurs in cattle feeds in the United States, Europe and South Africa, with contamination rates ranging from 5 to 19 percent (Krytenburg *et al.*, 1998).

Listeria monocytogenes tends to occur in poor-quality silages and big-bale silage. When grass is ensiled under anaerobic conditions, the low pH regime ensures that *Listeria* is excluded from the resulting silage. However, in big-bale silage a degree of aerobic fermentation may occur, raising pH levels and allowing the growth of *Listeria*. These bacteria also survive at low temperatures and in silages with high levels of dry matter. Contamination of silage with *Listeria* is important as it causes abortion, meningitis, encephalitis and septicaemia in animals and humans. The incidence of various forms of listeriosis has been increasing in recent years.

4. FUNGAL CONTAMINANTS

There are consistent reports of worldwide contamination of feeds with fungi and their spores. In the tropics, *Aspergillus* is the predominant genus in dairy and other feeds (Dhand, Joshi and Jand, 1998). Other species include *Penicillium*, *Fusarium* and *Alternaria*, which are also important contaminants of cereal grains (D'Mello, Macdonald and Cochrane, 1993). Fungal contamination is undesirable because of the potential for mycotoxin production (see next section). However, spores from mouldy hay, silage, brewers' grain and sugar-beet pulp may be inhaled or consumed by animals with deleterious effects termed "mycosis". Common examples of such conditions include ringworm and mycotic abortion. The latter may occur in cattle as a result of systemic transmission and subsequent proliferation in placental and foetal tissues.

5. MYCOTOXINS

Mycotoxins are those secondary metabolites of fungi that have the capacity to impair animal health and productivity (D'Mello and Macdonald, 1998). The diverse effects precipitated by these compounds are conventionally considered under the generic term "mycotoxicosis", and include distinct syndromes as well as non-specific conditions. A list of the principal mycotoxins occurring in feeds and forages is given in Table 1, which also indicates the fungal species associated with the production of these contaminants. Mycotoxin contamination of forages and cereals frequently occurs in the field following infection of plants with particular pathogenic fungi or with symbiotic endophytes. Contamination may also occur during processing and storage of harvested products and feed whenever environmental conditions are appropriate for spoilage fungi. Moisture content and ambient temperature are key determinants of fungal colonization and mycotoxin production. It is conventional to subdivide toxigenic fungi into "field" (or plant-pathogenic) and "storage" (or saprophytic/spoilage) organisms. *Claviceps*, *Neotyphodium*, *Fusarium* and *Alternaria* are classical representatives of field fungi while *Aspergillus* and *Penicillium* exemplify storage organisms. Mycotoxigenic species may be further distinguished on the basis of geographical prevalence, reflecting specific environmental requirements for growth and secondary metabolism. Thus, *Aspergillus flavus*, *A. parasiticus* and *A. ochraceus* readily proliferate under warm, humid conditions, while *Penicillium expansum* and *P. verrucosum* are essentially temperate fungi. Consequently, the *Aspergillus* mycotoxins predominate in plant products emanating from the tropics and other warm regions, while the *Penicillium* mycotoxins occur widely in temperate foods, particularly cereal grains. *Fusarium* fungi are more ubiquitous, but even this genus contains toxigenic species that are almost exclusively associated with cereals from warm countries.

An emerging feature is the co-production of two or more mycotoxins by the same species of fungus (Table 1). This observation has enabled a fresh interpretation of the causes of well-known cases recorded in the history of mycotoxicoses.

TABLE 1
Origin of principal mycotoxins occurring in common feeds and forage

Mycotoxins	Fungal species
Aflatoxins	*Aspergillus flavus; A. parasiticus*
Cyclopiazonic acid	*A. flavus*
Ochratoxin A	*A. ochraceus; Penicillium viridicatum; P. cyclopium*
Citrinin	*P. citrinum; P. expansum*
Patulin	*P. expansum*
Citreoviridin	*P. citreo-viride*
Deoxynivalenol	*Fusarium culmorum; F. graminearum*
T-2 toxin	*F. sporotrichioides; F. poae*
Diacetoxyscirpenol	*F. sporotrichioides; F. graminearum; F. poae*
Zearalenone	*F. culmorum; F. graminearum; F. sporotrichioides*
Fumonisins; moniliformin; fusaric acid	*F. moniliforme*
Tenuazonic acid; alternariol; alternariol methyl ether; altenuene	*Alternaria alternata*
Ergopeptine alkaloids	*Neotyphodium coenophialum*
Lolitrem alkaloids	*N. lolii*
Ergot alkaloids	*Claviceps purpurea*
Phomopsins	*Phomopsis leptostromiformis*
Sporidesmin A	*Pithomyces chartarum*

Aflatoxins

This group includes aflatoxin B1, B2, G1 and G2 (AFB1, AFB2, AFG1 and AFG2, respectively). In addition, aflatoxin M1 (AFM1) has been identified in the milk of dairy cows consuming AFB1-contaminated feeds. The aflatoxigenic *Aspergilli* are generally regarded as storage fungi, proliferating under conditions of relatively high moisture/humidity and temperature. Aflatoxin contamination is, therefore, almost exclusively confined to tropical feeds such as oilseed by-products derived from groundnuts, cottonseed and palm kernel. Aflatoxin contamination of maize is also an important problem in warm humid regions where *A. flavus* may infect the crop prior to harvest and remain viable during storage.

Surveillance of animal feeds for aflatoxins is an ongoing issue, owing to their diverse forms of toxicity and also because of legislation in developed countries (D'Mello and Macdonald, 1998). In the United

Kingdom, analysis conducted during the 1987-1990 period indicated that all imported feedstuffs complied with legislation in force for AFB1 levels. Elsewhere, however, aflatoxin levels in certain feeds still pose serious risks to animal health. Thus, in India total aflatoxin levels of 3 700 _g/kg were detected in a sample of groundnut cake. Of potentially greater significance is the contamination of maize samples in China and northern Viet Nam with combinations of AFB1 and *Fusarium* mycotoxins. In China, 85 percent of maize samples were contaminated with both AFB1 and fumonisin B1 at levels ranging from 8 to 68_g/kg and 160 to 25 970_g/kg, respectively. Feed-grade maize in northern Viet Nam had AFB1 levels ranging from 9 to 96 _g/kg, and fumonisin B1 levels in the range of 271 to 3 447 g/kg (Placinta, D'Mello and Macdonald, 1999). Between 1988 and 1989, analyses of farmgate milk in the United Kingdom showed low levels of AFM1 contamination, but more than 50 percent of milk samples in the United Republic of Tanzania were found to contain the mycotoxin (D'Mello and Macdonald, 1998). The importance of aflatoxins in animal health emerged in 1960, following an incident in the United Kingdom in which 100 000 turkey poults died from acute necrosis of the liver and hyperplasia of the bile duct ("turkey X disease"), attributed to the consumption of groundnuts infected with *Aspergillus flavus*. This event marked a defining point in the history of mycotoxicoses, leading to the discovery of the aflatoxins. Subsequent studies showed that aflatoxins are acutely toxic to ducklings, but ruminants are more resistant. However, the major impetus arose from epidemiological evidence linking chronic aflatoxin exposure with the incidence of cancer in humans.

Ochratoxins

The *Aspergillus* genus includes a species (*A. ochraceus*) that produces ochratoxins, a property it shares with at least two *Penicillium* species. Ochratoxin A (OA) and ochratoxin B are two forms that occur naturally as contaminants, with OA being more ubiquitous, occurring predominantly in cereal grains and in the tissues of animals reared on contaminated feed. Another mycotoxin, citrinin, often co-occurs with ochratoxin. In recent Bulgarian wheat samples, OA and citrinin levels ranged from < 0.5 to 39_g/kg and from < 5 to 420_g/kg, respectively. In oats, higher levels of OA were detected (maximizing at 140_g/kg) while citrinin was below detection limits (D'Mello, 2001).

The ochratoxins and citrinin are nephrotoxic to a wide range of animal species. OA is frequently implicated in porcine nephropathy and in Balkan endemic nephropathy of humans. The role of citrinin in these syndromes has yet to be elucidated.

Fusarium mycotoxins

Extensive data now exist to indicate the global scale of contamination of cereal grains and animal feed with *Fusarium* mycotoxins (D'Mello and Macdonald, 1998). Of particular importance are the trichothecenes, zearalenone (ZEN) and the fumonisins. The trichothecenes are subdivided into four basic groups, with types A and B being the most important. Type A trichothecenes include T-2 toxin, HT-2 toxin, neosolaniol and diacetoxyscirpenol (DAS). Type B trichothecenes include deoxynivalenol (DON, also known as vomitoxin), nivalenol and fusarenon-X. The production of the two types of trichothecenes is characteristic for a particular *Fusarium* species. However, a common feature of the secondary metabolism of these fungi is their ability to synthesize ZEN which, consequently, occurs as a co-contaminant with certain trichothecenes. The fumonisins are synthesized by another distinct group of *Fusarium* species (Table 1). Three members of this group (fumonisins B1, B2 and B3) often occur together in maize.

Virtually all the toxigenic species of *Fusarium* listed in Table 1 are also major pathogens of cereal plants, causing diseases such as head blight in wheat and barley and ear rot in maize. Harvested grain from diseased crops is therefore likely to be contaminated with the appropriate mycotoxins, and this is supported by ample evidence. Surveillance of grain and animal feed for the occurrence of *Fusarium* mycotoxins has been the subject of many investigations over recent years (Tables 2 and 3). The global distribution of these mycotoxins

TABLE 2
Global distribution of deoxynivalenol (DON), nivalenol (NIV) and zearalenone (ZEN) in cereal grains and animal feed (mg/kg)

Country	Cereal/feed type	DON	NIV	ZEN
Germany	Wheat	0.004-20.5	0.003-0.032	0.001-8.04
Poland	Wheat	2.0-40.0	0.01	0.01-2.0
	Maize kernels	4.0-320.0		
	Maize cobs: axial stems	9.0-927.0		
Finland	Feeds and grains	0.007-0.3		0.022-0.095
	Oats	1.3-2.6		
Norway	Wheat	0.45-4.3	max 0.054	
	Barley	2.2-13.33	max 0.77	
	Oats	7.2-62.05	max 0.67	
Netherlands	Wheat	0.020-0.231	0.007-0.203	0.002-0.174
	Barley	0.004-0.152	0.030-0.145	0.004-0.009
	Oats	0.056-0.147	0.017-0.039	0.016-0.029
	Rye	0.008-0.384	0.010-0.034	0.011
South Africa	Cereals/animal feed		0.05-8.0	
Philippines	Maize		0.018-0.102	0.059-0.505
Thailand	Maize			0.923
Korea, Republic	Barley	0.005-0.361	0.005-0.361	
	Maize	mean 0.145	mean 0.168	
Viet Nam	Maize powder	1.53-6.51	0.78-1.95	
China	Maize	0.49-3.10	0.6	
Japan	Wheat	0.03-1.28	0.04-1.22	0.002-0.025
	Barley			0.010-0.658
	Wheat	0.029-11.7	0.01-4.4	0.053-0.51
	Barley	61.0-71.0	14.0-26.0	11.0-15.0
New Zealand	Maize	max 3.4-8.5	max 4.4-7.0	max 2.7-10.5
USA	Wheat	up to 9.3		
	Wheat (winter), 1991	< 0.1-4.9		
	Wheat (spring), 1991	< 0.1-0.9		
	Wheat, 1993	< 0.5-18.0		
	Barley, 1993	< 0.5-26.0		
Canada	Wheat (hard)	0.01-10.5		
	Wheat (soft, winter)	0.01-5.67		
	Wheat (soft, spring)	0.01-1.51		
	Maize	0.02-4.09		
	Animal feeds	0.013-0.2	0.065-0.311	
[illegible]	Wheat	0.10-9.25		

Source: Adapted from Placinta, D'Mello and Macdonald, 1999.

TABLE 3
Worldwide contamination of maize and animal feeds with fumonisins (g/kg)

Country	FB1	FB2	FB3	Total
Maize				
Benin	nd1-2 630	nd-680		nd-3 310
Botswana	35-255	nd-75	nd-30	35-305
Mozambique	240-295	75-110	25-50	340-395
South Africa	60-70	nd	nd	60-70
South Africa	max 2 000			
Malawi	nd-115	nd-30	nd	nd-135
Zambia	20-1 420	nd-290		20-1 710
Zimbabwe	55-1 910	nd-620	nd-205	55-2 735
Tanzania, United Republic	nd-160	nd-60	nd	nd-225
Honduras	68-6 555			
Argentina	85-8 791	nd-11 300	nd-3 537	85-16 760
Uruguay	nd-3 688			
Costa Rica	1 700-4 780			
Italy	10-2 330	nd-520		10-2 850
Portugal	90-3 370	nd-1 080		90-4 450
Viet Nam	268-1 516	155-401	101-268	524-2 185
China	160-25 970	160-6 770	110-4 130	430-36 870
Philippines	57-1 820	58-1 210		
Thailand	63-18 800	50-1 400		
Indonesia	226-1 780	231-556		
Animal feed				
South Africa	4 000-11 000			
Uruguay	256-6 342			
India	20-260			

1nd = not detectable.
Source: Adapted from Placinta, D'Mello and Macdonald, 1999.

is a salient feature, but striking regional differences should also be noted. Another aspect worthy of comment is consistent evidence of the co-occurrence of various *Fusarium* mycotoxins in the same sample. These issues have been considered at greater length by Placinta, D'Mello and Macdonald (1999) who, for example, referred to a German study in which 94 percent of wheat samples analysed were contaminated by between two and six *Fusarium* mycotoxins and 20 percent of the samples were co-contaminated with DON and ZEN (Table 2). The most frequent combination included

DON, 3-ADON and ZEN. T-2 and HT-2 toxins were detected at levels ranging from 0.003 to 0.250 mg/kg and 0.003 to 0.020 mg/kg, respectively, but these mycotoxins only occurred in combination with DON, NIV and ZEN.

In the Lublin region of southeastern Poland, type A trichothecene contamination of barley grain was linked with the natural incidence of *Fusarium* head blight, in which the predominating organism was *F. sporotrichioides* (Placinta, D'Mello and Macdonald, 1999). Of 24 barley grain samples, 50 percent were positive for T-2 toxin, with a range of 0.02 to 2.4 mg/kg. In five of these samples, co-contamination with HT-2 toxin occurred, with a range of 0.01 to 0.37 mg/kg. Maize ears may also become naturally infected with *Fusarium* pathogens. The findings of one study in Poland indicated that infection with *F. graminearum* can result in contamination of cobs with DON (Table 2) and 15-ADON simultaneously (Placinta, D'Mello and Macdonald, 1999). Concentrations of DON and 15-ADON in Fusarium-damaged kernels ranged from 4 to 320 mg/kg and 3 to 86 mg/kg, respectively, but the axial stems of the cobs were more heavily contaminated, at 9 to 927 mg/kg (Table 2) and 6 to 606 mg/kg, respectively. Oat grains produced in Norway by commercial growers were found to be more heavily contaminated with DON than barley or wheat kernels (Table 2). In addition to NIV (Table 2), other contaminants included 3-ADON and fusarenon-X. For example, 56 percent of certain oat samples contained detectable quantities of 3-ADON at 0.03 mg/kg or more. Other notable examples of DON contamination include wheat and barley samples from Japan and the United States (Table 2). It should be stated, however, that even in samples with lower levels of contamination, high incidence rates have been recorded. Thus, 90 and 79 percent of cereal samples in the Netherlands were positive for DON and NIV respectively (Placinta, D'Mello and Macdonald, 1999).

Widespread contamination of maize and animal feed with fumonisins has recently been reported (Table 3). In most instances the predominant fumonisin was FB1. The highest values for FB1 were recorded from maize samples in China, where AFB1 co-occurred in

85 percent of samples, and in Thailand. Multiple contamination of maize with fumonisins, DON, NIV and AFB1 was also observed in northern Viet Nam. For FB2, the highest values in maize were found in samples from Argentina. In the Philippines, Thailand and Indonesia, FB1 and FB2 occurred in more than 50 percent of maize samples, and these mycotoxins co-occurred with aflatoxins in 48 percent of samples (Placinta, D'Mello and Macdonald, 1999).

The *Fusarium* mycotoxins induce a wide range of effects in farm livestock (D'Mello, 2000). DON is a potent feed intake inhibitor in pigs; ZEN is associated with reproductive abnormalities in pigs and ruminants. Fumonisins have been linked with specific syndromes, namely porcine pulmonary oedema and equine leukoencephalomalacia. Fumonisin contamination of maize in South Africa has been correlated with the occurrence of oesophageal cancer in humans.

Endophyte alkaloids

The endophytic fungus *Neotyphodium coenophialum* occurs in close association with perennial tall fescue, while another related fungus, *N. lolii*, may be present in perennial ryegrass (D'Mello, 2000). Ergopeptine alkaloids, mainly ergovaline, occur in *N. coenophialum*-infected tall fescue, while the indole isoprenoid lolitrem alkaloids, particularly lolitrem B, are found in *N. lolii*-infected perennial ryegrass. The ergopeptine alkaloids reduce growth, reproductive performance and milk production in cattle, while the lolitrem compounds induce neurological effects in ruminants.

Phomopsins

In Australia, lupin stubble is valued as fodder for sheep, but infection with the fungus *Phomopsis leptostromiformis* is a major limiting factor because of toxicity arising from the production of phomopsins by the fungus. Mature or senescing parts of the plant, including stems, pods and seeds, are particularly prone to infection. Phomopsin A is considered to be the primary toxin, causing effects such as ill-thrift, liver damage, photosensitization and reduced reproductive performance in sheep (D'Mello and Macdonald, 1998).

Sporidesmin

Pithomyces chartarum is a ubiquitous saprophyte of pastures and has the capacity to synthesize sporidesmin A, a compound causing facial eczema and liver damage in sheep.

6. PLANT TOXINS

Many plant components have the potential to precipitate adverse effects on the productivity of farm livestock (D'Mello, 2000). These compounds are present in the foliage and/or seeds of virtually every plant that is used in practical feeding. Typical concentrations for selected toxins are presented in Table 4. Plant toxins may be divided into a heat-labile group, comprising lectins, proteinase inhibitors and cyanogens, which are sensitive to standard processing temperatures,

TABLE 4
Plant toxins: sources and concentrations

Toxin	Principal sources	Typical concentrations
Lectins	Jackbean	73 units/mg protein
	Winged bean	40-320 units/mg
	Lima beans	59 units/mg protein
Trypsin inhibitors	Soybean	88 units/mg
Antigenic proteins	Soybean	-
Cyanogens	Cassava root	186 mg HCN/kg
Condensed tannins	*Acacia* spp.	65 g/kg
	Lotus spp.	30-40 g/kg
Quinolizidine alkaloids	Lupin	10-20 g/kg
Glusosinolates	Rapeseed	100 mmol/kg
Gossypol	Cottonseed	0.6-12 g/kg (free)
Saponins (steroidal)	*Brachiaria decumbens; Panicum* spp.	-
S-methyl cysteine sulphoxide	Kale	40-60 g/kg
Mimosine	*Leucaena leucocephala*	145 g/kg (seed) 25 g/kg (leaf)
Phyto-oestrogens	Clover; lucerne; soybean	-

Source: Compiled from D'Mello, 1995.

and a heat-stable group including, among many others, antigenic proteins, condensed tannins, quinolizidine alkaloids, glucosinolates, gossypol, saponins, the non-protein amino acids S-methyl cysteine sulphoxide and mimosine, and phyto-oestrogens. The role of these substances as antinutritional factors has been considered at length by D'Mello (2000), but the salient points are worth reiterating.

Lectins

Lectins are proteins capable of damaging the intestinal mucosa. In contrast to most other dietary proteins, lectins resist digestive breakdown and substantial quantities of ingested lectins may be recovered intact from the faeces of animals fed diets containing one of a number of legume seeds (D'Mello, 2000). The prime example of a lectin with potent antinutritional and toxic properties is concanavalin A, a component of the jack bean. Lectins are also present in other legume grains including the winged bean and soybean. Concanavalin A enhances the shedding of brush-border membranes and decreases villus length, thereby reducing surface area for absorption in the small intestine. With other lectins, the lamina propria of the intestine may become infiltrated with eosinophils and lymphocytes. The overall effect is reduced nutrient absorption, but immune function may also be impaired.

Proteinase inhibitors

The proteinase inhibitors are typical examples of heat-labile factors with antinutritional activity. They constitute a unique class of proteins with the ability to react in a highly specific manner with a number of proteolytic enzymes in the digestive secretions of animals. The trypsin inhibitors of soybean (Table 4) are now well characterized (D'Mello, 1995) and are important determinants of nutritive value. Proteinase inhibitors are also present in other leguminous seeds such as field beans, winged beans, pigeon pea and cowpea. Effects in animals include reduced protein digestion and endogenous loss of amino acids, with the overall result that performance is impaired.

Antigenic proteins

Certain storage proteins of legume seeds are capable of crossing the epithelial barrier of the intestinal mucosa to elicit adverse effects on immune function in farm animals. In the case of the soybean, the antigenic proteins have been identified as glycinin and _-conglycinin. The antigenic proteins are characterized by their resistance to denaturation by conventional thermal processing procedures and to enzyme attack in the digestive tract of mammals. The most striking effects of antigenic proteins are embodied within the "immune hypersensitivity" syndrome. This condition occurs after feeding heated soybean to sensitized calves and piglets (D'Mello, 1991). The component antigens provoke extensive local and systemic immunological reactions together with severe intestinal damage. The resulting effects include abnormalities in movement of digesta, impaired nutrient absorption and a predisposition to diarrhoea.

Cyanogens

Cyanogens occur widely in plants and in diverse forms. In sorghum and cassava (Table 4), the predominant cyanogens are, respectively, dhurrin and linamarin. The latter compound is also present in linseed. Cyanogens are glycosides that readily yield HCN and it is this latter molecule that causes dysfunction of the central nervous system, respiratory failure and cardiac arrest (D'Mello, 2000). Metabolizable energy values for poultry tend to be lower in untreated cassava root meal, presumably because of its cyanogenic potential.

Condensed tannins

Tannins belong to a group of phenolic compounds with a molecular weight in excess of 500 daltons. Condensed tannins (CTs) are a subset of this group and are widely distributed in leguminous forages (Table 4) and seeds and in sorghum. Cattle and sheep are sensitive to CTs, while goats are more resistant. Adverse effects may be seen in sheep when CTs, including those in lotus or in browse legumes such as *Acacia* species, comprise a significant part of their diets. Primary effects include impaired rumen function and depressed intake, wool growth and live-weight gain. However, at moderate levels (30 to 40g/kg legume dry matter), CTs may result in nutritional advantages in respect of increased bypass protein availability and bloat suppression in cattle. At higher levels (100 to 120 g CTs/kg legume dry matter), reduced gastrointestinal parasitism in lambs has been reported (D'Mello, 2000).

Quinolizidine alkaloids

The quinolizidine alkaloids occur in lupins and include lupinine, sparteine and lupanine. Bitter cultivars contain relatively high levels of total alkaloids (Table 4) and are not suitable as animal feedstuffs because of their negative effects on intake. In addition, cattle consuming certain lupin species during pregnancy may produce calves with multiple congenital deformities.

Glucosinolates

Glucosinolates are glycosides of particular significance in brassica forage crops such as kale (Table 4; D'Mello, 2000). Removal of glucose from glucosinolates by plant or microbial enzymes (myrosinase), results in the release of a diverse array of compounds which undergo further breakdown to yield a number of toxic metabolites. The most common breakdown products are isothiocyanates and nitriles but, depending on such conditions as pH, temperature and metallic ion concentrations, a number of other metabolites may also be produced. These products may then cause organ damage, goitrogenic effects or reduced feed intake, particularly in non-ruminant animals.

Gossypol

Gossypol pigment occurs in cottonseed (Table 4) in free and bound forms. In whole seeds, gossypol exists essentially in the free form, but variable amounts may bind with protein during processing to yield inactive forms. Free gossypol is the toxic entity and causes organ damage, cardiac failure and death. Cottonseed meal fed to bulls can induce increased sperm abnormalities and decreased sperm production.

Saponins

Saponins are divided into two groups: steroidal saponins, which occur as glycosides in certain pasture plants such as *Brachiaria decumbens* and *Panicum species* (Table 4); and triterpenoid saponins, which occur in soybean and alfalfa. Many hepatogenous photosensitization conditions in sheep have been attributed to the intake of forage plants containing steroidal saponins. In contrast, triterpenoid saponins from alfalfa reduce feed degradation in the rumen.

Amino acids

A wide range of non-protein amino acids occur in the foliage and seeds of plants. Forage and root brassica crops contain S-methyl cysteine sulphoxide (SMCO), while the aromatic amino acid mimosine occurs in the foliage and seeds of the tropical legume *Leucaena leucocephala* (D'Mello, 2000). Uncontrolled feeding of brassica forage to ruminants causes organ damage with haemolytic anaemia, which is attributed to the intake of SMCO. Abrupt feeding of *Leucaena* to sheep causes shedding of fleece, reduced intake, organ damage and death. In cattle, loss of hair, excessive salivation, lethargy, weight loss and enlarged thyroids are common features of *Leucaena* toxicity.

Phyto-oestrogens

Phyto-oestrogens are a diverse group of isoflavonoid compounds found primarily in forage and grain legumes (Table 4). In clover, formononetin is the major form of phyto-oestrogen. Phyto-oestrogens are actively metabolized in the rumen to form products that vary in their biological activity. Formononetin is converted into a more oestrogenic compound. Phyto-oestrogens have been associated with "clover disease" in sheep, which is characterized by low ovulation and conception rates (D'Mello, 2000).

7. WEED SEEDS

Contamination of animal feeds with weed seeds is a major problem worldwide. The impact of weed seeds arises from the toxins they contain and from their diluent effects on nutrient density of feeds. The toxins include many of those cited in the previous section, particularly alkaloids, saponins, amino acids and proteinase inhibitors. Examples of weed seeds that are controlled by legislation in various countries include those of *Datura* spp., common vetch, castor-oil plant and *Crotalaria* spp.

8. UNDECLARED ADDITIVES

Animal products are frequently contaminated with drug residues administered through the feed. Such feed additives may be used for disease control and the enhancement of livestock performance. Residues may also arise through contamination of animal feeds with undeclared drugs. The occurrence of these drugs is mostly due to cross-contamination in feed mills (Lynas *et al.*, 1998). For example, medicated feed residues may be retained within equipment and then contaminate subsequent batches of feed. Under these conditions, levels of contamination may be low but sufficient to cause detectable residues in animal products. Lynas *et al.* (1998) examined the extent of feed contamination with undeclared antimicrobial additives in Northern Ireland. Of 247 medicated feeds, 35 percent were found to contain undeclared antimicrobials; and of 161 unmedicated feeds, 44 percent were shown to contain antimicrobials. The contaminants most frequently identified included chlortetracycline, sulphonamides, penicillin and ionophores. Sulphadimidine in contaminated feeds was sufficient to cause violative tissue residues if consumed by animals in the finishing stages. It is possible that feed contamination with undeclared antimicrobials is a global problem warranting further investigation. Drug residues in animal products are undesirable because of human health implications concerning allergies and the development of antibiotic resistance in disease organisms.

9. REGULATION

It is instructive and relevant to provide a brief review of the regulatory prospects for the control of undesirable substances. Currently, regulations are most comprehensive in Europe and North America, while in developing countries statutory directives may not even exist. Thus, 50 countries, mostly in Africa, have no regulations for mycotoxin control (D'Mello and Macdonald, 1998). This situation may be changed by the new rules imposed on feeds imported into the European Union (EU) which came into force in August 1999. Non-EU feed manufacturers are now required to have representatives based in the EU who can confirm declarations concerning certain quality and safety standards for imported animal feeds.

For heavy metals and aflatoxins, distinctions in prescribed limits are generally made for straight, complete and complementary feedstuffs. Additional distinctions may apply according to the destination of feeds for a particular class of animal. For controlled pesticides, separate regulations exist for feedstuffs and for fats, with virtually no distinction for the class of animal. Of the wide variety of plant toxins, only gossypol, cyanogens and certain glucosinolates are subject to regulatory control in the EU. Special regulations apply to contamination of feeds with specific bacteria. For example, under United Kingdom regulations, positive identification of *Salmonella* in feeds must be reported to a "veterinary officer of the Minister" (HMSO, 1989).

10. EFFECTS OF PROCESSING

Heat processing is a common procedure in feed manufacture, conferring improved properties as regards the safety and nutritive value of animal feeds. For example, heat treatment of dried poultry litter appears to be an effective method for controlling, or even eliminating, contamination with *Salmonella*, *E. coli* and *Campylobacter* (Jeffrey *et al.*, 1998). Thermal processing is also effective for denaturing proteinase inhibitors, lectins and cyanogens. However, for antigenic proteins, more complex procedures involving the use of hot aqueous ethanol extraction is required (D'Mello, 1991).

For aflatoxin-contaminated oilseeds destined for animal feed, ammoniation appears to be the processing method of choice. The feedstuff is treated with either ammonium hydroxide or gaseous ammonia at high temperatures and pressure in commercial feed mills, or at ambient temperature and low pressure in small-scale operations in developing countries. If the ammoniation reactions are allowed to proceed to completion, the detoxification process is irreversible and aflatoxin contamination is virtually eliminated. Providing that the residual ammonia is dissipated, diets containing the decontaminated meals are readily consumed by animals without ill-effects. Depending on the efficacy of decontamination, residues of AFM1 in the milk of dairy cows are substantially reduced or absent altogether. The adverse effects of tannin-rich forages may be overcome by treatment or spraying of foliage with polyethylene glycol, but the practical application of this procedure still has to be economically evaluated.

11. CONCLUSIONS

Animal feed, including herbage, may be contaminated with organic and inorganic compounds as well as with particulates. Organic chemicals comprise the largest group and include plant toxins, mycotoxins, antibiotics, prion proteins and pesticides. Inorganic compounds include heavy metals and radionuclides. Particulates such as weed seeds and certain bacterial pathogens are common contaminants of feed. The effects of feed contaminants and toxins range from reduced intake to reproductive dysfunction and increased incidence of bacterial diseases. Residues transferred to edible animal products represent another reason for concern. Comprehensive legislation is in place for the control of several of these chemical compounds and pathogens in feed. However, in many developing countries, particularly in Africa, statutory control of contaminants is at best rudimentary. The scope for decontamination of feeds is limited and generally uneconomic, and prevention is the most effective practical strategy.

REFERENCES

Dhand, N.K., Joshi, D.V. & Jand, S.K. 1998. Fungal contaminants of dairy feed and their toxigenicity. *Indian Journal of Animal Sciences*, 68: 1095–1096.

D'Mello, J.P.F. 1991. Antigenic proteins. In J.P.F. D'Mello, C.M. Duffus & J.H. Duffus, eds. *Toxic substances in crop plants*, pp. 107–125. Cambridge, UK, Royal Society of Chemistry.

D'Mello, J.P.F. 1995. Antinutritional substances in legume seeds. *In* J.P.F. D'Mello and C. Devendra, eds. *Tropical legumes in animal nutrition*, pp. 135–172. Wallingford, UK, CAB International.

D'Mello, J.P.F. 2000. Antinutritional factors and mycotoxins. *In* J.P.F. D'Mello, ed. *Farm animal metabolism and nutrition*, pp. 383–403. Wallingford, UK, CAB International.

D'Mello, J.P.F. 2001. Mycotoxins in plant products. *In* J.P.F. D'Mello, ed. *Food safety*. Wallingford, UK, CABI Publishing. (in press)

D'Mello, J.P.F., Macdonald, A.M.C. & Cochrane, M.P. 1993. A preliminary study of the potential for mycotoxin production in barley grain. *Aspects of Applied Biology*, 36: 375–382.

D'Mello, J.P.F. & Macdonald, A.M.C. 1998. Fungal toxins as disease elicitors. *In* J. Rose, ed. *Environmental toxicology: current developments*, pp. 253–289. Amsterdam, the Netherlands, Gordon and Breach Science Publishers.

HMSO. 1989. Statutory instruments. *The Zoonoses Order* 1989. London, Her Majesty's Stationery Office (HMSO).

Jeffrey, J.S., Kirk, J.H., Atwill, E.R. & Cullor, J.S. 1998. Prevalence of selected microbial pathogens in processed poultry waste used as dairy cattle feed. *Poultry Science*, 77: 808–811.

Krytenburg, D.S., Hancock, D.D., Rice, D.H., Besser, T.E., Gay, C.C. & Gay, J.M. 1998. A pilot survey of *Salmonella enterica* contamination of cattle feeds in the Pacific northwestern USA. *Animal Feed Science and Technology*, 75: 75–79.

Lynas, L., Currie, D., McCaughey, W.J., McEvoy, J.D.G. & Kennedy, D.G. 1998. Contamination of animal feedingstuffs with undeclared antimicrobial additives. *Food Additives and Contaminants*, 15: 162–170.

Lynn, T.V., Hancock, D.D., Besser, T.E., Harrison, J.H., Rice, D.H., Stewart, N.T. & Rowan, L.L. 1998. The occurrence and replication of *Escherichia coli* in cattle feeds. *Journal of Dairy Science*, 81: 1102–1108.

MAFF. 1994. *Radionuclides in foods*. Food Surveillance Paper No. 43. London, United Kingdom Ministry of Agriculture, Fisheries and Food (MAFF), Her Majesty's Stationery Office.

Placinta, C.M., D'Mello, J.P.F. & Macdonald, A.M.C. 1999. A review of worldwide contamination of cereal grains and animal feed with *Fusarium* mycotoxins. *Animal Feed Science and Technology*, 78: 21–37.

van Barneveld, R.J. 1999. Physical and chemical contaminants in grains used in livestock feed. *Australian Journal of Agricultural Research*, 50: 807–823.

Antibiotic growth-promoters in food animals

Peter Hughes
Division of Microbiology, School of Biochemistry and Molecular Biology, University of Leeds, United Kingdom
John Heritage
Division of Microbiology, School of Biochemistry and Molecular Biology, University of Leeds, United Kingdom

There has been a developing controversy surrounding the use of antibiotics as growth promoters for food animals. These drugs are used at low doses in animal feeds and are considered to improve the quality of the product, with a lower percentage of fat and higher protein content in the meat. Other benefits of the use of antibiotic growth-promoters include control of zoonotic pathogens such as Salmonella, Campylobacter, Escherichia coli and enterococci. Use of any antibiotic is associated with the selection of resistance in pathogenic bacteria and it has been argued that the use of antibiotic growth-promoters imposes a selection pressure for bacteria that are resistant to antibiotics that may be used in clinical or veterinary practice, thus compromising the continued use of antimicrobial chemotherapy. This paper considers the use of antibiotics as growth promoters and then examines some of the alternative methods for achieving meat of high quality

1. INTRODUCTION

The term "antibiotic growth promoter" is used to describe any medicine that destroys or inhibits bacteria and is administered at a low, sub therapeutic dose. The use of antibiotics for growth promotion has arisen with the intensification of livestock farming. Infectious agents reduce the yield of farmed food animals and, to control these, the administration of sub-therapeutic antibiotics and antimicrobial agents has been shown to be effective. The use of growth-promoters is largely a problem of intensive farming methods and the problems caused by their use are largely those of developed rather than developing countries.

According to the National Office of Animal Health (NOAH, 2001), antibiotic growth promoters are used to "help growing animals digest their food more efficiently, get maximum benefit from it and allow them to develop into strong and healthy individuals". Although the mechanism underpinning their action is unclear, it is believed that the antibiotics suppress sensitive populations of bacteria in the intestines. It has been estimated that as much as 6 percent of the net energy in the pig diet could be lost due to microbial fermentation in the intestine (Jensen, 1998). If the microbial population could be better controlled, it is possible that the lost energy could be diverted to growth.

Thomke & Elwinger (1998) hypothesize that cytokines released during the immune response may also stimulate the release of catabolic hormones, which would reduce muscle mass. Therefore a reduction in gastrointestinal infections would result in the subsequent increase in muscle weight. Whatever the mechanism of action is, the result of the use of growth promoters is an improvement in daily growth rates between 1 and 10 percent resulting in meat of a better quality, with less fat and increased protein content. There can be no doubt that growth promoters are effective; Prescott & Baggot (1993), however, showed that the effects of growth promoters were much more noticeable in sick animals and those housed in cramped, unhygienic conditions.

Currently, there is controversy surrounding the use of growth promoters for animals destined for meat production, as overuse of any antibiotic over a period of time may lead to the local bacterial populations becoming resistant to the antibiotic. This is it not an invariable rule: *Streptococcus pyogenes* remains sensitive to penicillins after over sixty years of clinical use but such examples are, however, very rare. Undoubtedly, the medical exploitation of antimicrobial chemotherapy, particularly to treat human infections, has imposed an enormous selection pressure on formerly sensitive bacteria to acquire genetic elements that code for resistance to antibiotics.

This phenomenon has occurred within our hospitals, where compromised patients and over-use of antibiotics create an optimal environment for promoting resistance in susceptible strains of bacteria. One of the best examples of this is the methicillin-resistant *Staphylococcus aureus* (MRSA). Most strains of *S. aureus* produce penicillinases, a class of _-lactamase enzymes that break down penicillins, and so penicillinase-stable _-lactams, such as methicillin, cloxacillin and flucloxacillin, have been used to treat staphylococcal infections. The side chain on these molecules is of such a shape as to preclude binding of the _-lactamase to the antibiotic, thus rendering the drug stable. The widespread use of these drugs in the antibiotic arms race led to the emergence of strains of *Staphylococcus aureus* that resist _-lactamase through an alternative mechanism. MRSA have also acquired resistance determinants to a wide range of antibiotics (Hospital Infection Society, 2001). MRSA isolates often are resistant to a range of commonly used antimicrobial agents, including erythromycin, clindamycin, tetracycline and many aminoglycosides. The only therapeutic options for serious, systemic MRSA infection currently available are antibiotics in the glycopeptide family, including vancomycin. Since 1996, there have been reports of MRSA strains with decreased susceptibility to vancomycin (Centers for Disease Control and Prevention, 2001). Even more worrying is the observation that MRSA can express high-level resistance to glycopeptides when the vanA gene cassette is experimentally transferred from vancomycin-resistant enterococci (Noble, 1997). Were this to happen outside of the laboratory, this would take us one step closer to a post-antibiotic era where certain bacterial infections can no longer be treated with antimicrobial chemotherapy.

To avoid entering a post-antibiotic era, agencies around the world are examining our use and abuse of antimicrobials (House of Lords 1998; Commission of the European Communities, 2001).

Most antibiotics, around 60 percent, are used for therapeutic purposes in humans, although an increasing amount is administered as prophylaxis to prevent infections. The farming industry is the second largest consumer of antibiotics after medical practitioners. About 40 percent of antibiotics are used as growth promoters although antibiotics are also used therapeutically for animals. To reduce the risk of selecting resistant bacteria, the use of antibiotics must be restricted. The most attractive area for reducing the use of antibiotics is to ban their use as growth promoters in food animals. This review examines the consequences of the use of antibiotics as growth promoters and looks at alternatives aimed at reducing the pressure for the selection of resistance in bacteria that cause disease in both humans and animals.

2. CURRENT USE OF ANTIBIOTIC GROWTH PROMOTERS

On a world scale, the use of antibiotics as animal growth promoters differs dramatically. Sweden now makes no use of antibiotics for growth promotion purposes; the USA uses a wide range of antibiotics, including some considered to be "medically important". The following information is taken from the Report of the Joint Expert Advisory Committee on Antibiotic Resistance (JETACAR, 1999) on the use of antibiotics in food producing animals. Pigs are exposed to the greatest range of growth promoters. In the USA, for example, pigs are exposed to _-lactam antibiotics, including penicillins, lincosamides and macrolides, including erythromycin and tetracyclines. All these groups have members that are used to treat infections in humans. Pigs in the USA are exposed to a range of other compounds intended for growth promotion. These include bacitracin, flavophospholipol, pleuromutilins, quinoxalines, virginiamycin and arsenical compounds. In the USA, compounds used as growth promoters for cattle include flavophospholipol and virginiamycin, both also used as growth promoters in poultry. Cattle are also exposed to ionophores such as monensin to promote growth. Poultry are given arsenical compounds. The Animal Health Institute of America (AHI, 1995) has estimated that, without the use of growth promoting antibiotics, the USA would require an additional 452 million chickens, 23 million more cattle and 12 million more pigs to reach the levels of production attained by the current practices.

In Australia a range of growth promoters are employed. Pig farmers use arsenical compounds, flavophospholipol, the macrolides kitasamycin and tylosin, the quinoxaline olaquindox, and also virginiamycin, a streptogramin. Poultry producers use arsenical compounds, flavophospholipol, bacitracin and virginiamycin. Australian cattle farmers employ a range of ionophores, namely lasalocid, monensin, narasin and salinomycin. They also employ flavophospholipol and the macrolide oleandomycin. The glycopeptide avoparcin was withdrawn from the Australian market in December 1999 and existing stocks were not permitted to be used after 1 July 2000 when the registration lapsed. Use of this compound is discussed more fully below.

The use of growth promoters in the European Union (EU) is more limited. The oligosaccharide avilamycin is used in pig and poultry farming, ionophores, namely monensin and salinomycin are used for cattle and pigs and flavophospholipol is used with a range of livestock, including cattle, pigs, poultry and rabbits. In pig production, feed conversion efficiency is improved, along with daily growth rates, by approximately 2.5 percent. Mortality rates, associated with scouring and proliferative enteritis, are 10-15 percent lower than in countries, such as Sweden, who do not use antimicrobial growth promoters. In poultry, growth promoters, such as bacitracin, virginiamycin and avoparcin, control *Clostridium perfringens* infections, which are potentially fatal, in addition to improving feed conversion efficiency. It is estimated that this translated into an improvement of 1.5 percent, with added economic benefits from the reduction of *C. perfringens* infections (JETACAR, 1999).

The cattle industry in the USA is, perhaps, the most dependent on growth promoters as cattle have energy requirements that are high and that cannot be met easily without the use of growth promoters. High energy rations increase muscle growth and fat deposition in beef cattle, and help to improve milk productivity in dairy cattle. Unfortunately, the use of such rations is associated with side-effects, such as bloat and lactic acidosis, which can be debilitating or even fatal. These conditions are not a problem in Europe, where cattle diets contain more forage. To counteract this, monensin is used and, in addition to preventing the aforementioned conditions, it also significantly reduces ammonia and methane emissions (Mbanzamihigo *et al.*, 1995). It does not belong to a class of medically important antibiotics and is not associated with any major resistance problems. In an investigation to discover the long-term effects of monensin therapy, Rogers *et al.* (1997) concluded that there is no real adaptation of rumen microbes af er a period of 96-146 days of monensin treatment and that most effects disappear within a few hours following monensin withdrawal.

In this sense, monensin is probably one of the safest and most effective antibiotic growth-promoters with regard to human and animal health and associated bacterial resistance problems. Virginiamycin is used for similar purposes, such as the prevention of acid lactosis in cattle and poultry, but use of this compound has led to the selection of bacteria that are resistant to its effects. It is related to pristinamicin and quinupristin, both of which are used in human medicine and so there are fears that its continued use may compromise human therapy. Its use as a growth promoter has now been banned in the EU (Butaye *et al.*, 2000).

3. HUMAN HEALTH AND THE CONSEQUENCES OF USING ANTIBIOTIC GROWTH PROMOTERS

Human health can either be affected directly through residues of an antibiotic in meat, which may cause side-effects, or indi ectly, through the selection of antibiotic resistance determinants that may spread to a human pathogen. A drug that illustrates both potential problems is chloramphenicol. Gassner & Wuethrich (1994) have demonstrated the presence of chloramphenicol metabolites in meat products and have concluded that a link with the presence of these antibiotic residues in meat and the occurrence of aplastic anaemia in humans cannot be ruled out. Banned for growth promotion use in America over a decade ago and in the EU since 1994, chloramphenicol remains a drug used in treating typhoid fever. Over-use in animal husbandry is believed to have led to an increase in resistance to the drug in bacteria of the genus *Salmonella*, including *Salmonella typhi*, the causative bacterium of typhoid. It should be noted that the link between the use of antibiotic growth-promoters and increasing resistance remains unproven and that typhoid rates of infection and cure have not changed significantly since the introduction of the ban in 1994. There were 153 cases of typhoid fever in England and Wales in 1999, compared with 227 in 1994 and 132 in 1991 (Public Health Laboratory Service, On-Line). An alternative explanation for the increase in resistance to chloramphenicol is its availability as an over-the-counter drug in developing nations. It is cheap and relatively easy to produce.

In general, the effect of antibiotic residues in meat is insignificant when compared with the issue of selection and amplification of antibiotic resistant strains of bacteria. Antibiotic resistance determinants selected in this manner may have various routes by which they may compromise the therapeutic use of antibiotics. Selection may occur in microbes that are pathogenic for humans. Alternatively, resistance may be selected in zoonotic bacteria that subsequently cause human disease. On another level, the resistance determinant may be selected in a bacterium that is a member of the commensal flora of the animal being fed a growth promoter If such a resistance determinant is mobilisable, it may subsequently transfer to human or animal pathogens. The consequences of selection of resistance can range from prolonged illness and side effects, due to the use of alternative, and possibly more toxic, drugs, to death, following complete treatment failure. Modern medicine has furnished us with a wealth of antibiotics but, as the MRSA example discussed above illustrated, alternatives are starting to run out. The four types of bacteria most commonly associated with resistance due to use are *Salmonella*, *Campylobacter*, *Escherichia coli* and the enterococci; these bacteria are likely to be transmitted frequently from animals to humans.

Salmonella sp.

Bacteria of the genus *Salmonella* are responsible for many human diseases. *Salmonella typhi* is the causative agent of the potentially lethal typhoid fever. This infection is exclusively of human origin, spread via human faecal contamination of food or water and has a high mortality rate if left untreated. As it is mainly spread through insanitary conditions, and from human carriers, it is not within the scope of this review. Other species of *Salmonella* are zoonotic and are usually acquired from contaminated food sources, such as poultry. They commonly cause gastroenteritis. Symptoms can range from mild diarrhoea and nausea to severe vomiting, fever and violent diarrhoea. Most infections are confined to the gut but with people at the extremes of the age range, *Salmonella* may cause invasive disease. Such cases can require hospitalisation due to extreme dehydration coupled with endotoxic shock and may even result in death. The most common causative agents of salmonellosis are *Salmonella enterica* var. Enteriditis and *Salmonella enterica* var. Typhimurium and these bacteria are often found as contaminants in poultry and eggs. Recent measures to control *Salmonella* infections have proved very success ul in controlling the incidence of infection, particularly by *Salmonella enterica* var. Enteriditis (Advisory Committee on the Microbiological Safety of Food, 2001).

Salmonella enterica var. Typhimurium can usually be treated with fluoroquinolones, chloramphenicol and ampicillin. Molbak *et al.* (1999), however, reported that isolates of *Salmonella enterica* var. Typhimurium DT104, which is already resistant to multiple drugs including ampicillin and tetracycline, was showing resistance to quinolones. Consequently, some infections caused by DT104 can be very difficult to treat.

In 1998, Bonner reported that, in 1983, an outbreak of food poisoning caused by a resistant strain of *Salmonella* was linked to hamburgers made from cattle fed with chlortetracycline. Spika *et al.* (1987) traced a chloramphenicol-resistant strain of *Salmonella enterica* var. Newport from beef burgers to herds that had been dosed with chloramphenicol. Tacket *et al.* (1985) reported an outbreak of multi-drug resistant *Salmonella enterica* var. Enteriditis following consumption of raw milk. All of these reports originated in North America but this is not just an American problem; bacteria do not respect national boundaries. *Salmonella enterica* var. Typhimurium DT104 was isolated in the UK in 1988 and has subsequently migrated around the world.

Using medically important drugs, such as chloramphenicol and tetracycline, as growth promoters would seem to be the most obvious route towards resistant strains that pose a threat to human health, but the selection of resistance is not necessarily that simple. JETACAR (1999) report a significant correlation between the use of the aminoglycoside apramycin as a growth promoter and the isolation of resistant *Salmonella*, especially *Salmonella enterica* var. Typhimurium DT104, in cattle. Aminoglycoside resistance in these bacteria is due to the acquisition of n-acetylating enzyme. This enzyme also confers resistance to gentamicin, an important drug in human medicine.

Some scientists believe that antibiotic growth-promoters may contribute indirectly to the spread of *Salmonella* by allowing animals to be kept in unhygienic conditions. Growth promoters may act as masking agents for proper sanitation by reducing the pathogen load. The fundamentally unhygienic conditions of intensive broiler chicken production have been criticised. Broilers are reared in confined housing: this allows any pathogen to spread through the cohort rapidly.

Campylobacter sp.

Campylobacter, particularly *Campylobacter jejuni* and *C. coli*, is the most common cause of bacterial food poisoning in developed countries, such as the UK and USA. The Public Health Laboratory Service in the UK reported 53,858 faecal reports of *Campylobacter* in the year 2000, compared with only 14,844 faecal reports of *Salmonella* (Public Health Laboratory Service, On-Line). Gastrointestinal disease caused by *Campylobacter* shares many of the clinical symptoms of *Salmonella* infection, including diarrhoea, vomiting and fever. Hospitalisation is uncommon, affecting mostly those individuals at the extremes of age; death due to campylobacter infection is even less common.

C. jejuni, the commonest species to cause of human campylobacter infection, is sensitive to a range of agents, including erythromycin, chloramphenicol, tetracyclines, aminoglycosides and quinolones. Unusually for a Gram-negative bacterium, the agent of choice in infections requiring therapy is the macrolide erythromycin, which is also used as a growth promoter for pigs in America. This use is not believed to have compromised therapy. Due to the relative unimportance of *Campylobacter* in animal health, however, few studies on the resistance of animal isolates have been completed. The widespread use of macrolides in the food industry is of concern because of the clinical importance of this family of antibiotics. Thus, investigations into the occurrence of antibiotic resistance determinants in animal isolates are required.

If growth-promoting macrolide antibiotics have had little or no effect, then the therapeutic use of fluoroquinolones has had been associated with increased resistance. Fluoroquinolone-resistant strains are emerging around the world. Engberg *et al.* (2001) reviewed *in vitro* macrolide and quinolone resistance prevalence and trends in *Campylobacter* isolated from humans, showing a temporal relationship between use of quinolones in food animals and resistant isolates in humans. Endtz *et al.* (1991) reported that the use of fluoroquinolones to treat respiratory diseases in poultry seems to have led to the development of fluoroquinolone-resistant *Campylobacter* in the gut of treated birds. It is worth noting that the use of therapeutic drugs is often at doses that are orders of magnitude larger than the dose administered for growth promotion purposes.

Such a high dose exerts a huge selection pressure and, in the case of *Campylobacter*, has allowed resistant strains to emerge and even dominate.

Escherichia coli

When they are located in the gut, *Escherichia coli* strains are regarded as non-pathogenic, Gram-negative members of the commensal flora of humans and animals. They are, however, the frequent cause of a variety of human infections. Pathogenic strains are most commonly associated with urinary tract infections but strains are also the cause of traveller's diarrhoea. This bacterium is also involved frequently in abdominal infection, such as perforated bowel or appendicitis. It is also one of the most common causes of septicaemia. Rarely, it is the causative agent of neonatal meningitis. In summary, *E. coli* is capable of causing problems at almost any site of the body.

Some strains, the most notorious being *E. coli* O157, produce Vero cytotoxins and are referred to as Vero-Toxigenic *Escherichia coli* (VTEC) strains. These bacteria may cause haemorrhagic colitis and about 5 percent of cases progress to the haemolytic uraemic syndrome, with a case fatality rate of about 10 percent. This is the major cause of acute renal failure for children in the UK. In adults, symptoms of haemolytic uraemic syndrome are seen along with neurological complications. The natural reservoir for VTEC strains is the gastrointestinal tract of cattle and possibly other domesticated animals and so these bacteria may be subject to selection pressure from antibiotic growth promoters.

Antibiotic resistance in *E. coli* is widespread globally, with agents such as the penicillin found to be of decreasing efficacy against it (Heritage *et al.*, 2001). LeClerc (1996) warns about the dangers of complacency by reporting high rates of mutation in *E. coli* O157, following observations that they could acquire resistance determinants easily by horizontal gene transfer. It was noted that this was a possible route via which antibiotic resistance, from a pool of environmental pathogens, could be conferred. Even if antibiotic growth-promoters were not directly targeted against the bacteria, it remains possible that strains of this bacterium may acquire resistance from the gut microflora of the food animal. For this reason, and for the sheer volume and severity of disease it causes, it would be sensible not to ignore *E. coli* when considering the risks associated with the use of antibiotic growth-promoters.

Enterococci

The *enterococci*, such as *Enterococcus faecalis* and *Enterococcus faecium*, are of increasing concern, since they cause illness and death, especially in severely compromised patients in hospitals. Unlike the bacteria discussed above, the enterococci are Gram-positive and so are susceptible to most of the antibiotics used as growth-promoters. (Over)-use of antibiotics in a clinical setting has resulted in the selection of multi-resistant enterococci, including vancomycin-resistant enterococci. Such isolates are resistant to all conventional systemic antimicrobial therapies. Vancomycin-resistant enterococci were first isolated in Europe in the mid-eighties but quickly spread to the USA. Edmond *et al.* (1996) found that patients with blood-borne infections caused by vancomycin-resistant enterococci had over double the mortality of patients who were infected with enterococci susceptible to amoxycillin and vancomycin.

It has been suggested that the use of antibiotic growth-promoters, in particular the drug avoparcin, has contributed to the emergence of vancomycin-resistant enterococci. Both drugs are glycopeptides and the van genes found in enterococci confer resistance to both drugs. Use of avoparcin as a growth promoter increases the selective pressure for resistance within the animal. There is th s a risk that, subsequently, resistant bacteria may colonise humans. These resistant bacteria have the potential to cause disease, either in the colonised host or after spreading to another, susceptible host, such as an immunocompromised patient. Alternatively, organisms passing through the gut on meat may be able to transfer resistance genes to the resident microflora, via mobile genetic elements including transposons and plasmids.

Although avoparcin is not used in human medicine its analogues, vancomycin and teicoplanin are. Khachatourians (1998) noted that vancomycin-resistant isolates from Denmark and Germany are cross-resistant to avoparcin, illustrating that avoparcin-resistant bacteria in animals pose a potential threat to humans. The work of Das *et al.* (1997) shows that this potential threat may be a reality. A truck driver suffered a compound fracture of the femur whilst working in a chicken packing factory. He later developed a localised infection at the wound site. A wound swab yielded a mixture of *Proteus sp.* and *Enterococcus faecalis*. The *Proteus* was sensitive to the cefuroxime, with which the patient was treated. Subsequent swabs revealed a pure culture of vancomycin-resistant *Enterococcus faecalis*. Resistance in this bacterium resulted from expression of the vanA cassette. The factory was investigated and, out of 21 samples taken from surfaces and chickens, eight contained vancomycin-resistant enterococci that carried the vanA cassette. As the patient did not carry VRE in his faeces and had not been hospitalised or taken antibiotics within a recent period, it was concluded that the patient most likely became infected at the factory. If this was the case, it demonstrates the potential for avoparcin used as an animal growth promoter to compromise human health.

Following molecular genetic analysis of vancomycin-resistant enterococci from the faeces of poultry, pigs and humans, McDonald *et al.* (1997) found that there was evidence of transmission of the Tn1546 transposon, which codes for resistance to glycopeptides, between farm animals and humans. Van den Bogaard *et al.* (1997) used similar analyses to examine the prevalence of vancomycin-resistant *Enterococcus faecium* in faeces from turkeys and those involved in their cultivation. They discovered that at one of the 47 turkey farms tested, the isolates from the turkeys were identical to those of the farmer. Polymerase Chain Reaction (PCR) amplification and analysis and Pulsed-Field Gel Electrophoretic studies on vanA and other cassettes showed identity between the animal and the human isolates. Critics of this study have pointed out, however, that there was no evidence that the farmer was colonised with the bacterium and that reproducible observations in this case are lacking. The impact of avoparcin on the resistance problem is difficult to evaluate, but as the evidence from various studies accumulates, it would appear that there exists a theoretical risk of selection o glycopeptide resistance through use of avoparcin as an animal growth promoter and that resistant bacteria selected in this manner may subsequently compromise human health.

The European Commission banned the use of avoparcin as a growth promoter on the grounds of unknown risk. Del Grosso *et al.* (2000) found that, after the ban, a decrease was observed in contamination of meat products by vancomycin-resistant enterococci. The reduction was statistically significant in poultry (from 18.8 percent to 9.6 percent) but not in pork products (from 9.7 percent to 6.9 percent), allowing the conclusion that avoparcin withdrawal has been successful in reducing VRE contamination in meat products.

There are those who argue that antibiotic growth-promoters cause resistance that can be transferred to humans, or that antibiotic growth-promoters pose a threat to human health. To support this argument, it has been said that evidence to demonstrate the threat is non-existent because of methodological inadequacies, lack of reproducibility and limited data comparability. Critics also point out that the greatest threat to the continuing use of antibiotics in human medicine comes from the use of these agents in the treatment of infections in humans. What cannot be dismissed, however, is the fact that antibiotics foster resistance and that antibiotic growth-promoters are strongly implicated as a probable selective agent. The European Commission no longer permits "medically important" antibiotics to be used as antibiotic growth-promoters, due to possible risks of compromise of therapy. However, this needs to be a global effort. As Fidler (1996) noted, bacteria do not respect international borders.

In the early 1970s, the UK banned the use of tetracycline and penicillin for growth promotional purposes, spurring other European countries to take the same precaution shortly after. In the mid 1970s, the Food and Drug Administration (FDA) proposed a similar ban in the USA, but Congress intervened and required the FDA to do more research before instituting a ban. Today, the European Commission, the World Health Organisation, the Centers for Disease Control and the American Public Health Association all support the immediate prohibition of antibiotic growth-promoters that are the same as, or closely related to, antibiotics used in humans. In March 1999, the Centre for Science in the Public Interest, the Environmental Defence Fund, and others petitioned the FDA to ban, for purposes of growth promotion, six antibiotics used in or related to those used in human medicine, including penicillin, tetracycline, erythromycin, lincomycin, tylosin, and virginiamycin. The FDA has recently launched a Task Force (FDA, 2001) to tackle the subject of the use of antimicrobials in agriculture but many politicians have greeted it with negativity. It is worth noting that the Framework Document simply laid out a program for assessing the risk of antimicrobials on human health. In itself, it did not epresent any form of legislation, yet still was met with many negative comments that threatened its success. The most common criticism was that it lacked evidence.

4. THE ALTERNATIVES TO ANTIBIOTIC GROWTH PROMOTERS

In considering phasing out or banning antibiotic growth promoters, the quality of any alternatives, either on the market, that could be developed or that are available illegally, must be assessed. Essentially, there are two main ways in which we can reduce our dependence on antibiotic use in animals. An obvious choice is the development of alternatives to antibiotics that work via similar mechanisms, promoting growth whilst enhancing the efficiency of feed conversion. A more difficult route would be to improve animal health. Growth promoters have been shown (Prescott & Baggot, 1993) to perform best when conditions are worst: i.e. when the animal is in poor health and the living conditions unhygienic. If their local environment is improved, with overcrowding reduced and infection control techniques introduced, then the actual need for growth promoters may be removed.

In-feed enzymes

In-feed enzymes are routinely added to pig and poultry feeds and work by helping to break down certain components of the feed, such as _-glucans, proteins and phytates, that the animal may have problems digesting. They are produced as fermentation products from fungi and bacteria and seem to only have a positive effect on the animal. Some ethicists, however, have argued that adding enzymes to animals merely shows that we think of them as "factory beasts". Apart from ethical objections, in-feed enzymes are very effective at maximising feed conversion efficiency and have few drawbacks. As a result, current research is focussed on improving the quality of existing enzymes; whilst broadening the range of feed ingredients that they may be used to digest. The Scientific Committee for Animal Nutrition (2001) concluded that conditions of use evaluated so far are acceptable as regards to consumers, users and animals.

Competitive exclusion products

Competitive exclusion products are in-feed microbes consisting of a variety of species of bacteria that are marketed as being "friendly". The mechanism of action is believed to be that, by allowing such bacteria to colonise the gastrointestinal tract, potential pathogens are prevented from colonising the gut and thus causing infection. This is the competitive exclusion principle. These products are often administered to newborn animals, especially poultry, to colonise the gastrointestinal tract and prevent *Salmonella* and *Campylobacter* infections. It is not known how effective the treatment is but it is believed to reduce diarrhoea and reduce levels of mortality. These products are also given to animals that have been treated therapeutically with antibiotics, to re-colonise a gut that may have been depopulated by the ant microbial action of the drugs.

Probiotics

Probiotics are similar to competitive exclusion products. They are believed to improve the overall health of an animal by improving the microbial balance in its gut. The way they work has not been established, although it has been hypothesized that their action can be summarised in three ways. The first proposal is a reiteration of the competitive exclusion principle: by colonising the gut in large numbers, the probiotic bacteria exclude pathogens and thus prevent them from causing infection. The second possibility is that they act as a stimulus for the immune system. As the immune system is engaged following exposure to probiotic bacteria, any hostile bacteria are also noticed, following increased surveillance by leukocytes, and thus potential pathogens are eliminated. The third suggestion proposes that probiotics have a strong, positive influence on intestinal metabolic activities, such as increased production of vitamin B12, bacteriocins, and propionic acid. Other mechanisms have been proposed but remain to be confirmed.

The problem with probiotics is the lack of evidence as to their mechanism of action and of the effects on host animals. Shahani *et al.* (1983) demonstrated that the growth of experimentally induced tumours could be inhibited in mice fed with fermented colostrums, but only in animals dosed before tumour growth started. Kato *et al.* (1985) confirmed these experiments, showing that intraperitoneal administration of *Lactobacillus casei* inhibited tumour growth. Following these experiments, it was suggested that *L. casei* had immunopotentiator properties similar to those of BCG: in effect, a vaccine to tumours and a stimulus to the immune system. Unfortunately, these results could not be replicated on farms. Most probiotics would not be administered via the intraperitoneal route on a working farm. Disturbingly, it was also discovered that some strains could be harmful: Sharpe *et al.* (1973) found that *L. casei* subspecies rhamnosus could produce endocarditis or abscesses in host animals.

For probiotics to be used as immunopotentiators, as many studies have suggested, there remain a number of questions that must be answered. What are the most active strains and do they have the potential to be pathogenic? What is the maximum dose? When and how should the probiotic be delivered? The easiest option is an in-feed delivery system.

Probiotics are effective in certain cases, notably in newborn animals or those that have been treated with antibiotics, where they have the same effect as competitive exclusion products. They may also be useful in helping to boost weight gain and feed conversion rates. Antibiotic growth-promoters have a lot of evidence, supported by excellent field results, to prove their efficacy. The usefulness of probiotics, however, remains unproven: they do have some strong supporters in the scientific community, but these are matched by an equal number of detractors. The beneficial effects of probiotics have been demonstrated, almost exclusively under defined experimental conditions. There have only been a small number of well-designed, double-blind, controlled trials to support the health-promoting claims of probiotic therapy.

An additional problem caused by the use of live bacterial products is that there may be potential dangers concerning antibiotic resistance and cryptic virulence factors. The Australian government is considering introducing legislation on probiotics, to require monitoring for the presence of resistance plasmids and the resistance patterns associated with antimicrobials of every strain of bacterium within a preparation, as part of the registration procedure. A recent report from the Scientific Committee for Animal Nutrition (2001) concerning the safety of a probiotic product found that two of the principal strains within the product, *Pediococcus acidilactici* and *Lactobacillus plantarum*, were resistant to tetracyclines. Resistance was found to be coded for by the tet (S) gene, which is often located on highly mobile genetic elements. As a result, it was concluded that because of the possible dissemination of tetracycline resistance genes in animal bacterial populations, the food chain and the environment, the use of that product poses a risk when used in animal nutrition.

Infection control measures

The use of antimicrobials as growth promoting agents rests on their role in controlling infection in growing animals. Similarly, many of the alternatives are aimed at controlling infection, often indirectly. But what of direct measures used to control infection in farm animals?

The Australian Pig Farming Industry pioneered the "all-in-all-out" method of pig production. This is a new system, used to replace the older technique of having a constant stream of pigs moving through the farm. Instead of having a range of ages, all the pigs weaned within a week are designated into a single cohort and are housed together in one shed. They are not allowed to mix with pigs from other cohorts and so cross-infection between groups is prevented. "Segregated early weaning" takes note of the observation that the sow is an important source of pathogens. If piglets are weaned early, they are less likely to come into contact with pathogens from their mothers. Care must be taken not to create welfare problems by weaning animals too early, however.

The "specific pathogen-free" system is used to prevent pigs from acquiring many of the diseases that require antibiotic intervention, especially respiratory disease. To achieve this, they are born by hysterectomy and hand reared. This will only be cost-effective for valuable breeding stock.

Finally, vaccination is used to offer protection against certain pathogens, such as enterotoxigenic *E. coli* and various mycoplasma infections. One of he major drawbacks to all these schemes is the huge cost involved. In Australia, many of the farms are very large, by European standards, so they can afford to implement these measures. Some UK farmers have started to adopt infection control methods, in particular "all-in-all-out" husbandry.

The Swedish model

Sweden posed the question of suitable alternatives to antibiotics in 1985, when its Parliament passed the Feedingstuffs Act and banned the use of antibiotics for growth promotion. Calves, turkeys and fattening pigs did not appear to be affected significantly by the ban: growth rates may have decreased slightly, but there were no major increases in mortality. Efforts were made to establish new feeds and housing for broiler chickens and, after an initial "unsettled" period of outbreaks of necrotic enteritis, were considered successful. Animal

Welfare groups were also happy to note that the new conditions were much better for the birds. Weaned pigs did not, however, enjoy the same level of success. JETACAR (1999) reported that scouring increased and there was an increase of 1.5 percent in mortality, roughly fifty thousand pigs. Furthermore, the time taken to reach 25kg increased by 5-6 days. Housing and hygiene were improved, livestock buildings were partitioned to help slow the spread of disease, "all-in-all-out" measures were introduced and the feed was changed. This resulted in a 50 percent reduction in the use of antibiotics by 1993 and was followed by further reductions each year. In 1995-96, only 11 percent of weaners were treated with antibiotics in their food.

The ban stimulated new ideas and technologies and successfully achieved its goal of an agricultural industry independent of growth promoting antibiotics. In addition, animal welfare was improved overall. The Swedish experience shows that antibiotics are not necessary to produce healthy animals, provided their living conditions, rearing and foods are improved. This did come at a cost: thousands of pigs and chickens probably died as a direct result of the ban, despite the overall improvement in animal welfare. Swedish produce is more expensive and so less competitive on the market, and the costs of the venture are expensive. It may be argued, however, that the problems encountered by adopting the Swedish model have been justified by the outcome. Sweden has shown the rest of the world that it is possible to have modern farming without the use of antibiotics as growth promoters.

5. CONCLUSIONS

The best alternative to Antibiotic growth-promoters is a general improvement of conditions for animals that produce our food, following, for example, the Swedish model. Medically important antibiotics must be prohibited from use in a growth promotional role as a matter of immediacy. Unfortunately, reform can be slow and extremely costly, as Sweden showed. In order to start a reformation of the Industry as a whole, it's essential that attitudes to the use of antibiotic growth promoters be changed. While the greatest threat to the continued use of antibiotics comes from human medicine, selection of resistance is a problem that affects everyone. It matters very little to someone whose antibiotic treatment is failing if selection of the resistant strain resulted from clinical overuse of the antibiotic or from other sources.

ACKNOWLEDGEMENTS

The authors would like to thank Prof. Mike Forbes of the University of Leeds for his critical reading of this paper and his valuable discussion.

REFERENCES

Advisory Committee on the Microbiological Safety of Food. 2001. *Second report on Salmonella in eggs*. Food Standards Agency, London.

Animal Health Institute (USA). 1998. Antibiotic resistance back in the news. *AHI Quarterly*, 19: 1–4.

Bonner, J. 1998. Hooked on Drugs. *New Scientist* http://archive.newscientist.com/ Last ac essed 28 October 2001.

Butaye P., van Damme, K., Devriese, L.A., van Damme, L., Bael, M., Lauwers, S., & Haesebrouck, F. 2000. *In vitro* susceptibility of *Enterococcus faecium* isolated from food to growth-promoting and therapeutic antibiotics. *International Journal of Food Microbiology*, 54: 181–187.

Centers for Disease Control. 2000. Issues in healthcare settings: laboratory detection of oxacillin/methicillin - resistant *Staphylococcus aureus* (MRSA). http://www.cdc.gov/ncidod/hip/Lab/FactSheet/mrsa.htm Last accessed 28 October 2001.

Commission of the European Communities DGXXIV Scientific Steering Committee. 1999. Opinion of the Scientific Steering Committee on Antimicrobial Resistance. http://europa.eu.int/comm/food/fs/sc/ssc/out50_en.pdf Last accessed 28 October 2001.

Das, I., Fraise, A. & Wise, R. 1997. Are glycopeptide-resistant enterococci in animals a threat to human beings? *The Lancet*, 9057: 997.

Del Grosso, M., Caprioli, A., Chinzari, P., Fontana, M.C., Pezzotti, G., Manfrin, A., di Giannatale, E., Goffredo, E. & Pantosti, A. 2000. Detection and characterization of vancomycin-resistant enterococci in farm animal and raw meat products in Italy. *Microbial Drug Resistance-Mechanisms Epidemiology and Disease*, 6: 313–318.

Edmond, M.B., Ober, J.F., Dawson, J.D., Weinbaum, D.L. & Wenzel, R.P. 1996. Vancomycin-resistant enterococcal bacteraemia: natural history and attributable mortality. *Clinical Infectious Diseases*, 23: 1234–1239.

Endtz, H. G. & Ruiijs, G. 1991. Quinolone resistance in *Campylobacter* isolated from Man and poultry following the introduction of fluoroquinolones in veterinary medicine. *Journal of Antimicrobial Chemotherapy*, 27: 199–208.

Engberg, J., Aarestrup, F.M., Taylor, D.E., Gerner-Smidt, P., & Nachamkin, I. 2001. Quinolone and macrolide resistance in *Campylobacter jejuni* and *Campylobacter coli* : resistance mechanisms and trends in human isolates. *Emerging Infectious Diseases*, 7: 24.

Fidler, D.P. 1996. Globalization, international law and emerging infectious diseases. *Emerging Infectious Diseases*, 2: 77–84.

Food and Drug Administration, USA. http://www.fda.gov/oc/antimicrobial/taskforce2000.html Last accessed 28 October 2001.

Gassner, B. & Wuethrich, A. 1994. Pharmacokinetic and toxicological aspects of the medication of beef-type calves with an oral formulation of chloramphenicol palmitate. *Journal of Veterinary Pharmacology and Therapeutics*, 17: 279–283.

Heritage, J, Ransome, N., Chambers, P.A. & Wilcox, M.H. 2001. A comparison of culture and PCR to determine the prevalence of ampicillin-resistant bacteria in the faecal flora of general practice patients. *Journal of Antimicrobial Chemotherapy*, 48: 287–289.

Hospital Infection Society. Revised guidelines on the control of MRSA in hospitals. htt ://www.his.org.uk/work/MRSA.html Last accessed 28 October 2001.

House of Lords Select Committee Report. 1998. *Resistance to antibiotics and other antimicrobial agents.* HMSO, London.

Jensen, B.B. 1998. The impact of feed additives on the microbial ecology of the gut in young pigs. *Journal of Animal and Feed Sciences,* 7: 45-64, Suppl. 1.

Joint Expert Advisory Committee on Antibiotic Resistance. 1999. Report of the Joint Expert Advisory Committee on Antibiotic Resistance (JETACAR) on the use of antibiotics in food producing animals: antibiotic resistant bacteria in animals and humans http://www.health.gov.au/pubs/jetacar.pdf Last accessed 28 October 2001.

Kato, I., Yokokura, T. & Mutai, M. 1985. Induction of tumoricidal peritoneal exudate cells by administration of *Lactobacillus casei. International Journal of Immuno-Pharmacology,* 7: 103109.

Khachatourians, G. 1998. Agricultural use of antibiotics and the evolution and transfer of antibiotic resistant bacteria. *Canadian Medical Association Journal,* 159: 1129–1136.

LeClerc, J.E. 1996. High mutation frequencies among *Escherichia coli* and *Salmonella* pathogens. *Science,* 274: 1208–1211.

Mbanzamihigo, L., Vannevel, C.J. & Demeyer, D.I. 1995. Adaptation of rumen fermentation to monensin administration. *Reproduction Nutrition Development,* 35: 353–365.

McDonald, L.C., Kuehnert, M.J., Tenover, F.C. & Jarvis, W.R. 1997. Vancomycin resistant enterococci outside the health care setting: prevalence, sources, and public health implications *Emerging Infectious Diseases,* 3: 311–317.

Molbak K. Baggesen D.L. Aarestrup F.M. Ebbesen J.M. Engberg J. Frydendahl K. Gerner-Smidt P. Petersen A.M. & Wegener HC. 1999. An outbreak of multidrug-resistant, quinolone-resistant *Salmonella enterica* serotype Typhimurium DT104. *New England Journal of Medicine,* 341: 1420–1455.

National Office of Ani al Health (NOAH). Antibiotics for animals. http://www.noah.co.uk/issues/antibiotics.htm Last accessed 28 October 2001.

Noble W.C. 1997. Antibiotic resistance in the staphylococci. *Science Progress,* 80: 5–20.

Prescott J.F. & Baggot J.D. 1993. *Antimicrobial Therapy in Veterinary Medicine*, 2nd edition, pp 564-565: Iowa State University Press.

Public Health Laboratory Service. http://www.phls.co.uk/facts/Gastro/Salmonella/salmtyphAnn.htm. Last accessed 28 October 2001.

Rogers, M., Jouany, J.P., Thivend, P. & Fontenot, J.P. 1997. The effects of short-term and long-term monensin supplementation, and its subsequent withdrawal on digestion in sheep. *Animal Feed Science and Technology*, 65: 113–127.

Scientific Committee for Animal Nutrition. 2001. Report of the Scientific Committee for Animal Nutrition (SCAN) on the Safety Assessment of Probiotic Product Pronifer MSB http://europa.eu.int/comm/food/fs/sc/s an/out58_en.pdf Last accessed 28 October 2001.

Scientific Committee for Animal Nutrition. 2001. Report of the Scientific Committee for Animal Nutrition (SCAN) on the use of certain enzymes in animal feeding stuffs. http://europa.eu.int/comm/food/fs/sc/scan/out52_en.pdf Last accessed 28 October 2001.

Shahani, K., Friend, B. & Bailey, P. 1983. Antitumor activity of fermented colostrum and milk. *Journal of Food Protection*, 46: 385-6.

Sharpe, M., Hill, R. & Lapage, S. 1973. Pathogenic lactobacilli. *Journal of Medical Microbiology*, 6: 281–286.

Spika, J.S., Waterman, S.H. & Soo Hoo, G.W. 1987. Chloramphenicol resistant *Salmonella* newport traced through hamburger to dairy farms *New England Journal of Medicine*, 316: 565–570.

Tacket, C.O., Dominguez, L.B., Fisher, H.J. & Cohen, M.L. 1985. An outbreak of multiple-drug-resistant *Salmonella enteritidis* from raw milk. *Journal of the American Medic l Association*, 253: 2058–2060.

Thomke, S. & Elwinger, K. 1998. Growth promotants in feeding pigs and poultry ii; mode of action of antibiotic growth promotants. *Annales de Zootechnie*, 47: 153–167.

Van den Bogaard, A.E.J.M., Jensen, L.B. & Stobberingh, E.E. 1997. Vancomycin-resistant enterococci in turkeys and farmers. *New England Journal of Medicine*, 337: 1558–1559